The End of the Future:
Governing Consequence
in the Age of Digital Sovereignty

Stephanie Polsky

The End of the Future:
Governing Consequence
in the Age of Digital Sovereignty

Stephanie Polsky

Academica Press
Washington – London

Library of Congress Cataloging-in-Publication Data

Names: Polsky, Stephanie, author.
Title: The end of the future : governing consequence in the age of digital sovereignty / Stephanie Polsky.
Description: Washington : Academica Press, 2019. | Includes bibliographical
references and index. | Summary: "In The End of the Future, Stephanie Polsky conceives an understanding of the digital through its dynamic intersection with the advent and development of the nation-state, race, colonization, navigational warfare, mercantilism, and capitalism, and the mathematical sciences over the past five centuries, the era during which the world became "modern." The book animates the twenty-first century as an era in which the screen has split off from itself and proliferated onto multiple surfaces, allowing an inverted image of totalitarianism to flash up and be altered to support our present condition of binary apperception. It progresses through a recognition of atomized political power, whose authority lies in the control not of the means of production, but of information, and in which digital media now serves to legitimize and promote a customized micropolitics of identity management"-- Provided by publisher.
Identifiers: LCCN 2019027587 | ISBN 9781680531572 (hardback) | ISBN
9781680531831 (paperback)
Subjects: LCSH: Technology and state. | Technology--Social aspects. | Technology and civilization. | Information technology--Political aspects.
Classification: LCC T49.5 .P69 2019 | DDC 303.48/3--dc23
LC record available at https://lccn.loc.gov/2019027587

Contents

Preamble

This book is first and foremost dedicated to my grandfather Eugene Polsky, who taught me from the earliest of ages that a life of the mind was the highest form of being. I developed as an independent thinker under his ardent tutelage from the youngest of ages. Throughout my formative years, he populated my imagination with the ghosts of many precious histories, establishing within me a desire to seek out their revival, even if the task ahead often seemed daunting. He was an avid reader of history and philosophy throughout his life. After I received my doctorate we began to call ourselves "The Drs Polsky." During my last phone call with him, he reluctantly admitted that he would be late with his "homework," meaning finishing a piece by Gilles Deleuze I had recently "assigned" to him. I sensed that he was indicating this was going to be the end of our term, and with a very heavy heart, I ended the call. He passed away within weeks of this conversation.

I heard the news through another telephone call received a few minutes shy of midnight standing in the late December cold at Waterloo Bridge about to ring in the new year. In many ways, that fateful hour has similarly never struck for me and so it was that my writing became a project to extend the expiry date of his vision and purpose. Shortly after his funeral, I dedicated myself to writing a critical trilogy dedicated to the concepts and concerns that motivated us both. I commenced this work starting with *Walter Benjamin's Transit: A Destructive Tour of Modernity*, continuing with *Ignoble Displacement: Dispossessed Capital in Neo-Dickensian London* and culminating with *The End of The Future: Governing Consequence in the Age of Digital Sovereignty*.

I first started my academic training in the hopeful age of the mid-1990s. Britain had shaken off Thatcher and many thought with it the odious spectre of neoliberalism. We never saw the maniacal potential of "New" Labour (though of course, it did what it said on the tin) until it was too late. That was until Tony Blair seated himself in the lap of George W. Bush, demonised the poor through antisocial behaviour orders and sold off the public housing stock and parts of the NHS to his rich liberal cronies. He is now, of course, fully recognised as an elder statesman of neoliberal capitalist ideology.

Few of us at the time were able to draw an explicit connection between the arrival of a Cool Britannia and a departure from the European Union. Similarly, no one could predict that a couple of decades later we would end up with a Brexit run by the aristocratic rentier class, which has not the capacity to assign meaning to Brexit, but rather to assign a generation to a new and novel means of serfdom, a strategically misguided minority of whom actually voted in favour of their own suffering.

This was not the only indignity this millennial generation would be subjected to at the hands of their establishment betters. During the post-crash austerity years of a Tory-Lib Dem coalition led government universities started to be complicit in acts that made their staff and students sense viscerally their own vulnerability to the market. Faculty started to train students to sell themselves and to choose intellectual projects that would merit funding if not stir their intellectual curiosity. Students were told to mimic the success of others rather than risk failure.

The last year before my US departure saw the maturation of that risk aversion reflected in Britain's inability to address crisis at any level. Soon after the Brexit vote, my classes became about allowing students the space for understanding, for support, and for the expression of humanity. It was very hard work to do in the world. It also was the best work I thought I could be doing. The university, at the time, was reaching a suicidal stage with European student numbers haemorrhaging, incidents of xenophobia on and around campus rising, European staff resigning, departmental budgets being slashed to the bone, and a return to quietism in the intellectual work being produced and disseminated in the curriculum. It was like being in a mental ward walking around the university corridors riddled with fear of coping with even less and whispers of staff cuts. It was time to go.

I came to the US thinking that this was the more stable academic market. I also knew intellectually this was the place to be at the dawn of the Western world's next great era of fascism. Nobody who wrote *Transit* could not realise, as Walter Benjamin did, the necessity of staying in Europe and reporting from inside its collapse as a civilisation.

What I underestimated landing in northern California was that I would find perhaps a more insidious evil to be at my doorstep in nearby Silicon Valley. At that time, we were all becoming aware of Facebook's complicity in Trump's election via Cambridge Analytica and Steve Bannon. Twitter was wilfully acting as Trump's official mouthpiece for what amounted to his daily barrage of hate speech and misinformation. Facebook's vice president of global public policy, Joel Kaplan, attended the sexual assault hearings of Brett Kavanaugh as a supporter. California, the supposedly most liberal place to land in Trump's America turned out in many ways to be emblematic of the racial, economic, sexual, and cultural divides that gave rise to it.

It was only then that I was seeing how this all worked in practice at ground level. I had no idea of the depth of things until I started to live up close to neoliberalism's most destructive features in the place where they prototype new and nefarious designs for its next consumption. Indeed, Silicon Valley's whole financial empire is built on the premise of individuals contributing to its data storehouses an infinite amount of unremunerated cognitive labour. This phenomenon is being closely tracked by contemporary sociologists, psychologists, anthropologists, economists, geographers, historians, and political scientists who have all in one respect or another professionally profited from the valuation of data. Moreover, these disciplines act as the perfect academic complement to Silicon Valley and thusly are richly rewarded with funding for their efforts within what remains of the academy. At the same time, humanities subjects such as ancient and modern languages, literature, law, philosophy, theology, and art are left to languish finding their funding reserves profoundly diminished.

Walter Benjamin warned us about using statistics in a fascist sense. The purpose of data generation is for humanity to be compelled towards greater complicity with forms of behaviour that extend the reach of neoliberal governance. This brings to mind Derrida's work on Hamlet which forces us to acknowledge that "the time is out of joint," and can only be restored when we realise justice itself as a futural act whose appearance will likely never be seen by any of us coerced to occupy the present as it is, rather than as it should be. In such a space, we must move as a plurality to avoid detection and to disorganise ourselves to achieve the effect of a haunting of the system.

Perhaps, it is time to revisit that decrepit site of moral dilemma we call time, history, the world, and to reckon what happened to all those dead futures that neoliberalism has rather conveniently terminated for us. I am reminded here, of Mark Fisher whose suicide clouded the entire last year I was at Goldsmiths, lingering as a spectre, a cypher for the misery all around us dwelling there in the very corridors of his profound despondency. If we need to resurrect Carthage then so be it, something is rotten there too. That said perhaps all roads lead to Rome, and indeed, to Carthage. It is only here that it is possible to sift through the possible legacies, the possible spirits, reaffirming one and not the other. Finally, in order to follow Walter Benjamin's posthumous direction, we must assume that we are the heirs of the future receptive of the foretelling of those who came before us.

Acknowledgements

Thanks to Dr Cecilia Wee my "agent" in both my work and my life, who never allows me to identify as anything other than Dr Polsky. Your constant support and unwavering confidence in my work have been invaluable. Your faith in me has acted as my north star on occasions where I was at a loss about my situation. You reassured me time and again that I was the person who needed to put this work out in the world. I am not sure I would have the desire to persevere in doing this writing if it were not for our shared mission of social justice and revolutionary education. As reluctant a Marxist as I am, I consider you my truest intellectual comrade, as well as my dearest friend.

Thanks to Rebecca Gribbon whose home has become my haven over many years and whose enthusiastic interest in my work helped produced the title of this work, "The End of the Future" over the course of an afternoon tea there. Thank you for making me feel as though I will always have a place in the world to return to whenever I am ready to do so. Our cherished friendship in many ways signifies that aspirational homecoming for me.

Thanks to members of the Data {Publics} Reading Group at Goldsmiths Magda Krysztoforska, Beverley Gadsden, Eleni Odysseos, and Anna Mikkola for inspiring me to push further and more forthrightly into this work through our lively discussions and your desire to learn from texts that, in sum, pieced together an unofficial history of algorithmic life over the course some eighty years. Our journey together was one of the fondest of my career and contributed directly to the ideas featured here. I hope that each of you will continue to find novel ways to deploy your critical thought in the world.

Thanks to Mohammad Mehdi Kimiagari my faithful reader and champion of my work in what were some of the darkest days of this project. I will always remember the gift of your imploring me back to my writing desk. I am very pleased to be able to support you as you forge ahead on your own path toward cultivating a life of the mind. I look forward to the flourishing of your work in the world.

Thanks to Christian Friess for being such an inventive and reflective office mate and friend during my time the Vienna University of Technology, and joining us as a late member of the Data {Publics} Reading Group. Your perspicacious insights pepper this text, and our lively and wondrous discussions via Vienna and California contributed both materially and virtually to the formulation of this project.

Thanks to Deirdre Daly, Aleardo Zanghellini, and Julie Bygraves whose contemplative viewpoints challenged and inspired me during the process of generating this work and whom I consider treasured political allies, as well as dear lifelong friends. Thanks to my cherished friend and sister Claire Collison for always lifting my hopes and spirit, guiding me to a higher plane of being in the world, as your Stephaniesque.

Introduction
Life on the Algorithmic Estate

We are now entering an era where the human world assumes recognition of itself as data. Its basis for existence is becoming fully subordinated to the software processes that tabulate, index, and sort the relations entailed in making up what we perceive as reality. The acceleration of data threatens to relinquish ephemeral modes of representation to ceaseless processes of computation. This situation compels the human world to form relations with non-human agencies, to establish exchanges with the algorithms and other software processes that accelerate and intensify the possibility of its obsolescence in order to allow for a profound upgrade to our own ontological understanding to take precedent. Through a partial attunement to what is always already non-human in its form of mediation to a higher intelligence, we are able to rediscover the actual inner logic of the age of intelligent machines, as at once the reason of trauma and the instrument of catastrophe for a humanity still beholden to a linear process of rationality.

Humanity now finds itself captive to pervasive institutionalised forms of violence whose force has everything to do with the aggressive pattern of economics in a neoliberal age, bound with an internet that has taken on a fourth dimension to generate consequence in the material world. This has allowed the internet to become thingly insofar as it will soon be understood not as an interface but as an environment. It thus takes on the ability to shape conditions beyond the imaginary and embed itself into materiality in a variety of ways that benefit from the demise of state parameters and the enlivenment of a fluidity of information able to migrate across time and space. All previous forms of media suffered from imprisonment within a screen, which limited their ability to function as the foundation for alternative networks, or as nodes of multilayered connectivity. The promise of a universal connectivity through a perpetual summons of our inclinations brings forth nothing less than a new form of imperialism able to transform space into a sphere of liquidity, and complexity into a condition of movement. As labour precarity and labour migration becomes the normative situation of a disenfranchised humanity, so too does its acquiescence to a universal accessibility where individuals are continuously subject to digital interpolation, and as such their behaviours and movements are made available to generating exploitable forms of interest. The data created is thus credited to others and interpreted to advance the interests of others; all in order to finance the states and corporations who control the means for making worlds we alone discover.

The prospect of a future that has the capacity to come to an end prompts us to apprehend an understanding of the digital through its dynamic intersection with the advent and development of the nation-state, race, colonisation, navigational warfare, mercantilism and capitalism, and the mathematical sciences over some five centuries. Its task is to animate an understanding of the twenty-first century as an era where the screen has split off from itself and proliferated onto multiple surfaces, and as a consequence has allowed an inverted image of totalitarianism to flash up at this point in history, and be altered to support our present condition of binary apperception. It progresses through a recognition of a now atomised political power whose authority lies in the control not of the means of production, but of information, and thus, digital media now serves to legitimise and promote a customised micropolitics of identity management. On this new apostolate plane, it is possible to conceive a world in which each human soul is captured and reproduced as an autonomous individual bearing affects and identities. The digital infrastructure of the twenty-first century makes it possible for power to operate through an esoteric mathematical means, and for factual material to be manipulated in the interest of advancing the means of control. This pathway towards understanding the significance of the digital travels a wide course between Elizabethan England, North American slavery, German fascism, Cold War cybernetic social engineering and counterinsurgency, and the (neo)libertarianism of Silicon Valley in order to arrive at a place where a cool organising intelligence that started from an ambition to resourcefully manipulate bodies, ends with their profound neutralisation.

The digital in this sense is made to denude its relationship to historical political economies, and to relate to ecologies of culture and media production spanning centuries as opposed to mere decades. In that time span, it has emerged variously as the quintessence of political representation, the essence of public perception and the immaterial layer through which the impacts on everyday life might be felt. The digital has managed to relate collective values to individual identities, and in so doing draw up an inventive terrain all its own in which these positionings are now given the power to become operational and indeed operative when it comes to our understanding of subjectivity itself. The digital has become essential to humanity's self-understanding through its ability to implicate itself within a pattern of changing societal and perceptual hierarchies that exists everywhere around it and from which it is now possible to conclude have assumed their own 'digital consciousness'. Over time, the digital comes to be tied to an understanding of emerging institutional protocols and new forms of non-representational visualities as they emerge in various guises within the algorithmic estate.

It is possible to witness this through the evolution of various public spheres whose spatial coordinates now encompass the dynamic realities of both a material and digital world. A diversity of critical perspectives including those drawn from visual culture, media studies, cultural studies, science and technology studies, ethnography, cultural theory, critical race theory, postcolonial theory, environmental humanities, and sociology together provide a compendium through which to grasp the historical consequence of data. These critical perspectives when brought together present a compelling new narrative around data as a technology of public truth and private architectures, aesthetically produced through new practices of capitalism. Within this challenging terrain, modes of information can be placed in further context and situated in conjunction with new and emerging regions of power. Here contemporary perception and cultural history can be joined together in various guises to generate a cogent understanding of data as both a vehicle for political agency, as well as its repression and, therefore, as an elemental force that figures crucially within any analysis of the turbulent realities of our time.

Such times and indeed the notion of time itself has become radically altered by the rise of digitisation so that memory itself has become the stuff of cheap storage and facile retrieval. We believe in a mode of recall now that is at once universal and progressively omniscient. The softness of software as we have come to know it presents humanity with the tantalising prospect of our bodies being subject to interiorised programmability as a means of optimising their functionality. It is by no means a coincidence that this is happening at a time humanity is grappling with its own biological finitude in this world. The human bodies that exist amidst that anxiety are now further burdened in their effort to conserve their resources by a body of information that, by contrast, has no looming expiry date. As such the algorithm's solutions-based thinking may well be on a collision course with our best interests and its capacity for overtaking our authority positioned well beyond our current scope of projection.

In a tangible sense, 'programmability' already fuels the current organization of the contemporary neoliberal state. Moreover, in many respects, computer code now functions as its own mechanism of causality and sovereignty. Similarly, the social, political, and economic coding embedded within our societies, now increasingly is materialised in the software and hardware that render us as conscious subjects, rather than the other way around. What we perceive has become the effect of technologies that are becoming increasingly adept not only in the prediction of our behaviour but in subtly influencing it. These technologies persuade us to accept a concept of the future that is very much furnished to us on the basis of past data. Our interface with the algorithm over time could easily become shorthand for everything we believe in, performing as a medium of interpretation of our lived reality. It is invisible yet generates visible, logical effects, and this holds vast implication for our perception of everything from genetics to the invisible hand of the market, from ideology to culture. A faith in the algorithm, more generally, is providing an esoteric bridge to an understanding between theoretical concepts and pragmatic reality. Data has generated a capacity of invocation that permits it to not merely describe the world, but to recast it in its own image. The algorithm in this way possesses and draws from a long lineage of alchemy, philosophy, and cybernetics. It does so, in order to both account for reality and direct it towards a future where humanity will be buffered by assisted intelligence, new media aesthetics, and Anthropocenic economics.

Digital technology will allow humanity to both anticipate its responses and schematise new routes through the algorithmic estate, as it itself relentlessly labours to reconfigure a method of accounting for reality that corresponds with new systems of programmable value. These complex configurations are likely to emerge as a crucial feature of ontological determination with a focus turned to gender, race, sexuality, and citizenship as its core problem sets. With the expansion of datafication so too comes an enlarged territory in which to exercise vanguard surveillance. Technologies will proliferate in this century which are poised to exert greater control over the meaning of information, the construction of the worlds we inhabit and increasingly the identities assigned to us externally in the cause of making us of 'value'. However, entities like Google, Facebook, and Amazon have limits to their practices and face from time to time temporal disruption. Indeed, if we accept surveillance as a series of technologies and practices that are informed by the long history of identity formation, it is possible to them place it within a greater narrative of sovereign security practices. As mass surveillance has long been and continues to be, a social and political norm so too is it possible to construe social formations such as race as a technology in its own right. Viewed within this context, classification emerges throughout the course of some three centuries, as a technology that ignites the possibility of capitalism to mature and for policing to emerge as its complementary function within the confines of the Anglo-American empire and beyond to the new Asian-American axes of global surveillance. The periphery between their development is a rich line through which to plumb the depths of what a codified world has yet to offer us by way of discursive and material practices seeking greater legitimate authority.

Categorical differentiations endure as the remainders of colonial processes and it is precisely through them that the concept of "the human" is and remains universalised through a hierarchy of subtle exclusion. In other words, "freedom" in its liberal form is only made possible through conditionality. At the same time, the peoples who create the conditions of possibility for that freedom must be assimilated or forgotten to maintain that standard. **The End of The Future** takes as one of its premises the analysis of liberalism within the context of the colonial state. The legacy of liberalism, therefore, can only be recognised when it is re-joined with what has been methodically separated from it in the past to secure its promotion and significance within the present political and economic context of neoliberalism. Through re-examining a limited series of western imperialism's pivotal events, it becomes possible to represent those absented matters and conceive a mode of reading history that is intimately drawn through the bodies that are its subjects. It is only here that a knowledge of self can be built from information; from the assignment of national boundaries, causal chronologies, and conceptions of history, politics, economics, and culture, and ultimately can be dissolved through a denial of those means.

This process allows us to recognise how those histories have so often been rendered as a dead past, rather than as an active and violating force in the world today. In terms of the future, there are expectations of how it should look and how it should be. What is missing and might be potentially revealed to us through colonial resonances is the uneven, recursive nature and qualities assigned to the future we endeavour to project for ourselves. The concerns of **The End of the Future** therefore ultimately lie with the advent of new imperial formations with the digital sphere, which come forth from the remnants of occluded histories, gradated sovereignties, and affective security regimes, in order to animate the scene of "new" racisms, bodily exposures, active debris, amidst the carceral archipelagos of colony and camp. These new formations carve out the distribution of inequities and deep fault lines of duress in new ways that complement the damaging political effects and commercial practices associated with the imperial projects of the past.

What distinguishes my research from that of my peers is the unique periodisation of this volume and how it views the rise of data and digitisation from the perspective of an imperial spatialisation of power. This book explores how technology over a period of centuries has come to mediate what it means to be human. It engages with forms of media and technology as they make their way across historical periods in order to determine and critically evaluate how such forms function as a part of a larger cultural, social, and political mapping. The book encourages readers to reconsider the development of older forms of media in light of today's technologies and to anticipate imminent possibilities for their further productive intersection. This book goes back to the mathematics of cryptography born of an even earlier medievalism to appreciate how the digital came to exemplify reality itself with aspirations towards its own language of transcendence. It also takes us back to the early modern beginnings of race as we understand it as a fractional measurement, and as a means of verifying human identity, and extrapolates that figuration onto the contour of a postwar cybernetics and its preoccupation with the apprehension of humanity's physical and behavioural characteristics.

This volume takes us from the relentless Nazi punch carding of difference to the impervious Cold War informatics of domination, towards today's contemporary genomic technology holding the promise of a micropolitical governance to come. This volume joins the founding of Silicon Valley to the territorial debasement of various geographies that found its rapidly increasing geopolitical predominance. The book probes the digital as a new planetary infrastructure that makes of a mathematical means an end that relies on the end of humanity as we have come to know it. The book concludes that on its present course, humanity as a subjective entity is aggressively being refashioned to conform with a re-engineered definition of it, whereby formulation becomes destiny and information becomes ontology.

The End of The Future will appeal to readers who are seeking an interdisciplinary understanding of the role data will play now and into the future in the evolution of human identity and its encoding as a complex web of pluralities. This book is dedicated to radically reshaping our understanding of how subjectivity is produced via information and distributed through the vast expanse of commercialised calculation and circulation. The way that data is being used to enact such understanding raises a great deal of concern related to issues of surveillance, privacy, security, profiteering, profiling, and social disaggregation amongst many others. This book intends to provide its readership with a cogent, multidisciplinary analysis of a palliated future produced through the netting of collective intelligence and the working of separable means over the course of some five centuries. This phenomenon is taking place on a global scale wherein a mollified image of authoritarianism secures its constituencies at the sacrifice of its imaginary adversaries.

The Stating of Emergency

This book is the third in a trilogy that began with *Walter Benjamin's Transit: A Destructive Tour of Modernity and continued with Ignoble Displacement: Dispossessed Capital in Neo-Dickensian London and now culminates with The End of The Future: Governing Consequence in the Age of Digital Sovereignty.* My starting point in writing has always been Walter Benjamin's essay "On the Concept of History" which quotes Flaubert: 'Few people can guess how despondent one has to be in order to resuscitate Carthage' (Benjamin, "On the Concept of History" 391) The nature of this lament becomes clear once we appreciate that the narrative of that society's total destruction comes to us via its vanquishers; the Romans. Therefore, Benjamin contended that there was no document of culture which is not at the same time a document of barbarism. The documentation of history itself bears this impression without exception and furthermore transmits its legacy to future generations. The only means open to us to counter that effect is to assume that what we are experiencing now as a living population is a version of Carthage. Our neo-imperial rulers share in common with Rome a predilection towards military aggression, the enactment of vengeful wars, and the wholesale seizure of native territories for later commercial exploitation. Rome's desire to utterly devastate Carthage was based not solely on its desire to enact what amounted to a genocidal campaign against the Carthaginian people, but equally to initiate terror amongst both Rome's enemies and allies. It is said that the Roman general Scipio Aemilianus wept when he ordered the destruction of the city and behaved virtuously toward the survivors. We shall never know as such stories are passed from one listener to another. For Benjamin that was of little importance. Instead, Benjamin wanted us to recognise that the oppression of the Carthaginians was not the exception, but rather the rule of life within the Roman empire.

We continue to live within a "state of emergency" that is keenly felt by those categorically oppressed by it, even as it remains obscure to those who occupy privileged positions endowed with social and cultural authority. Benjamin's critical works implored us to 'attain to a conception of history that accords with this insight' (392). The consequence of not raising the alarm as it were, 'to bring about a real state of emergency' was the furtherance of totalitarianism, and eventually, the advent of genocide (392). Benjamin cautioned, 'one reason fascism has a chance is that, in the name of progress, its opponents treat it as a historical norm' wherein society was progressing through aid of technology and that 'the working class was moving with that current' (392). Those opponents were German Marxists who 'preferred to cast the working class in the role of a redeemer of future generations, in this way cutting the sinews of its greatest strength. This indoctrination made the working class forget both its hatred and its spirit of sacrifice, for both, are nourished by the image of enslaved ancestors rather than by the ideal of liberated grandchildren' (394). The working classes were never going to be the victors over fascism, but rather its willing converts through such facile rhetoric, the consequence of which comes down to us to this day when we speak of the liberating potential of technology for the masses, rather than the source of their perpetuated enslavement.

We know that those who lived through the sacking of Carthage did not experience it as either homogeneous or empty time. Rather they were forced to imagine the disaster befalling them prismatically, making recourse to the clarification or distortion afforded by a particular viewpoint. Benjamin asserts that there is a secret agreement between past generations and the present one down through the order of remembrance. Remembrance brings historical time to a standstill, energetically repositioning it such that under the right circumstances it can take what Benjamin called a 'tiger's leap' into the future. The spirit of survival is what carries it unnaturally forth beyond worlds that have long since ceased. History, therefore, must be reconstructed as a mode of intervention, as a means of exhuming Carthage from the ceaseless mounding of events in which it would otherwise be interred.

In that same spirit, this present volume takes as its methodology an historical archaeology. As such, it does not aim for description or prescription, but rather for extraction and restoration. This technique, when applied to a body of knowledge, allows us to draw connections between our era and a series of very specific earlier ones in order to allow an underlying narrative to emerge. Benjamin believed that the most critical tool available to a culture was emulation. Through it comes the ambition to recapture and remember the subjectivity of the past and the ability for the past to reveal its enduring potential to shape the present. The aim of this book, like its predecessors, is to reveal the simultaneity of certain philosophies and ideologies that telescope through vast expanses of historical time.

Benjamin writes, 'Marx says that revolutions are the locomotives of world history. But perhaps it is quite otherwise. Perhaps revolutions are an attempt by the passengers on the train – namely the human race- to activate the emergency brake' (Paralipomena to "On the Concept of History" 402). One of the revolutionary aspects of Benjamin's analysis was the way in which it revealed how the concept of a homogenised, empty continuum of time contributed to the furtherance of a coercive and hierarchical society. In his critical work, he sought to create an alternative concept of history in an attempt to bring about a state of emergency that would break the spell of that deadly form of continuity with the past.

Although seldom acknowledged, Benjamin had an intellectual antipathy towards Marxism which caused him to have a subtly adversarial relationship with Bertolt Brecht. Equally, Benjamin was wary of the social research of Max Horkheimer and Theodor Adorno. They tried repeatedly to foist upon him their particular brand of dialectical Marxism in exchange for much needed financial support. The same can be said for Brecht. The intellectual price demanded for such support – allegiance to Marxism– was too high for Benjamin and therefore he forfeited his chance to become a legitimately recognised scholar in his time and an émigré to America. Benjamin dreaded the prospect of settling in the United States. In May of 1940, he wrote that he feared he was destined to be carted up and down the country, exhibited as the "last European", that is to say, the last of a generation of German Enlightenment thinkers ("Reflections" xiv). By September of that year, he would be dead.

Benjamin's enthusiasm for historical materialism died the previous year with the signing of the Hitler-Stalin pact. That was when he finally saw behind that curtain, that the game was rigged, that a system of mirrors gave the illusion of a transparency amongst the players that simply did not exist. 'One can imagine a philosophic counterpart to this apparatus…called "historical materialism". If it 'is to win all the time' it has to enlist the services of theology, which today, as we know, is small and ugly and has to keep out of sight' ("On the Concept of History" 389). From that point, Benjamin forced his mind to scan the murkiness of the past, rather than rely on the clear recognition of a present enemy, in order to wrest meaning from history. This is an incredibly fragile critical exercise mixing theology with metaphysics that few have attempted since, given what such archaeology extracted from the life force of its theoretical progenitor. His 'Theses' were found posthumously amongst his possessions. They are composed largely of quotations and related notes and it is probable that in this preliminary form they were never intended for publication. Benjamin was likely working on something larger to which these observations would contribute. Quotations, however, formed an active part of his intellectual methodology throughout his life and therefore cannot be overlooked. He employed quotation as a way to approach his subjects in order to insert the lever of his criticism into the subtlest aspects of their position. Quotation for Benjamin functioned as a diagnostic tool capable of restoring power, and therefore, criticism for him was a fundamentally recuperative act.

Interest in Benjamin's work was revived in the 1970s, by the famous art critic John Berger. Through his book *Ways of Seeing* and its subsequent television series, Berger brought Benjamin back from the brink of intellectual obscurity. *Ways of Seeing* became Berger's most famous work, the premise of which was basically a quotation of Benjamin's essay "Art in the Age of Technological Reproduction". Berger also was one to want to readily place Benjamin in the Marxist camp by suggesting that he was a card-carrying member of the Frankfort School. Not much of course was said about how little Benjamin was concerned with the class struggle as opposed to the epochal struggle of his times; the threat posed to German Enlightenment thinking by radical nationalist ideologues such as Heidegger and Jünger. Through close analysis of their reactionary metaphysical writings, Benjamin essentially clocked the regressive potential of counterculture decades before anybody else did. Benjamin was also one of the first of his generation to recognise technological mediation as a form of epistemology. He was one of the few thinkers to be able to inspect both closely and at a distance political systems within the context of their ability to condition and control our reality. Benjamin's montage reading of history takes place against a mid-twentieth century backdrop of political crisis wherein mass media had triumphed by transforming politics into an optical instrument and history into a projector of force.

Storage

Starting from the early 1970s, something inhuman has largely taken over cultural production. New recording technologies had ushered in a post-material era of perception. In 1971, Memorex emerged as the industry leader in recordable media. It launched itself into the public's imagination through an advertising campaign that asked audiences to distinguish if what they were hearing was live or a Memorex recording. The commercial campaign featured the famous recording artist Ella Fitzgerald who would sing a note that shattered a glass while being recorded to a Memorex audio cassette. The tape was played back and the recording also broke the glass, asking "Is it live, or is it Memorex?". This campaign ran well into the 1980s and suggested to audiences that technology, independent of humans, was now able to achieve material effects in the world that equalled their capacity. Memorex audio cassette recording was one of the last analogue technologies to dominate popular culture. Ella Fitzgerald achieved her fame during World War II and her career flourished through the early Cold War era. By the early 1970s she had become something of a cultural relic. Her appearance in the Memorex adverts suggested that technology itself was now able to surpass her natural abilities as an artist. At this time, the senses themselves were construed as part of a greater neural operating system that had virtually nothing to do with human involvement. Rather they took up residence in the mind through a series of pre-programmed steps at one with the brain's seamless powers of assimilation. The mind's autonomic quality enabled it to form impressions that lent an illusion of self-sameness and continuity to what was, in reality, a radical break with the past.

Analogue was giving way to digital and in this way time itself was subject to elongation. The great thing about the digital was that it didn't have to end. Digitisation solved an existential crisis first posed by the birth of mechanical reproduction. It put the death of the artist on a slow-motion collision course with information, such that social recognition was eventually rendered obsolete in favour of the requirements of the user. Artist production, in this way, became Bowie's cocaine, as opposed to Benjamin's hashish, an inhibitor of melancholy and a gateway drug to a more generalizable vacuity. The future would now come to humanity uninterrupted and the individual's uniquely human impressions of the world subject to infinite and arcane storage. Vietnam was the first political crisis beamed into our consciousness that could no longer be regarded as an event, but rather as a series of human failures to register any definitive form of understanding.

By the second half of 1970s political movements travelled not by underground, but rather 'On the waves of the air' ("Night Fever"). The Bee Gees track "Night Fever" from its beginning lyric recognises this is a uniquely conspiratorial, fugitive sort of social movement: 'If it's somethin' we can share, we can steal it' ("Night Fever"). What is apprehended in this fibril environment was a sense of a point of, or to the origin. Rather, popular culture was suddenly all about anonymous transmission. This was happening at a time when human memory was being evacuated, seized from a technological standpoint, rendering the individual unnecessary, while the 'we' was rendered as commercial equity.

According to the group's website, In the late 1950s, the Bee Gee's were originally The BG's - contrived from the common initials of their founders Barry Gibb, Bill Goode, and Bill Gates. The name then evolved from The BG's to the Bee Gees which eventually came to mean the Brothers Gibb. The term 'The Bee Gees' is not inherently subjective, nor does it differ significantly from person to person. By the early 1970s, the BG's moniker becomes a waste product of that history deemed antithetical to the fusion sounds of digital recording which required the brothers to literally fuse their output, recording their voices simultaneously so that all three vocal parts became inseparably intertwined in the mind of the listener. The Bee Gees soon after became an integral part of the social movement known as disco. 'Disco, in both avant-garde club and mainstream modes, was not interested in personal liberation; disco's aim was to reveal that memory is programmed by the source in advance and that the freedom is not "about" anything except the rules that gave rise to it' (Lin 95). Disco, therefore, becomes the perfect model of imitation and of iteration at a time where history as a medium is facing an intense period of collective evacuation. As a mode of pneumonic storage, it is profoundly open to appropriation, but not interpretation.

Disco emerged at a time of massive social upheaval and economic downturn, and yet in the mind of its listener, it reduced all possible concerns to feeling. In this way, culture became an ambient phenomenon, rather than the offspring of objective media. In so doing, it rendered the Frankfurt School's dialectical Marxism obsolete. By contrast, Benjamin's intellectual standard-bearer, the copy withstood the test of time. It had enabled the transfer of data to become synonymous with human action and enabled the human to increasingly resemble the machine. In the most basic protocols of behaviour, it was now difficult to distinguish the artist from the interpreter. As such creative output could thereafter be classified as a fundamentally diverse undertaking. At the same time, what is perhaps more troubling is that Benjamin's prediction was coming to pass: that avant-garde practices would dissolve into something generic, with the consequence that what was most politically irrelevant would pass into the slipstream of culture. As a consequence, there would be no collective sense of an emergency in need of breaking.

In the decades to follow, consciousness became progressively interiorized, essentially reversing the entire arrangement of society to track instead with mathematical operation, a formula capable of co-aligning with the feedback loop of probability that gave rise to it. History was now construed as nothing more than a repetitive 'set of effects' (Lin 95). In other words, as the Bee Gees crooned, 'We know how to do it.' The environmental portent of that dubious assumption in our century is what becomes suddenly critical. That we put out a recovery effort to revive the older analogue technology of memory, and in so doing, restore awareness to a humanity now fundamentally bereft of it. Buried within the collective consciousness lies an obscured history whose meaning, although corrupted through the processes of technological coercion, might eventually be rescued and restored in order to provide something of value ancillary to information, and thusly redeemable for some future constructive purpose. If realised, it would become the ultimate form of recycling.

In his last letter, Benjamin writes, 'my great fear is that we have much less time at our disposal than we imagined' ("Correspondence" 683). Therefore, he makes recourse to space to up his chances of beating the puppet, and by looking at the claims for freedom, justice, and equality that have gone unrecognised throughout history as an index of the value of history itself. Benjamin remarks, 'The past carries with it a secret index by which it is referred to redemption' ("On the Concept of History" 390). Redemption cannot achieve its span, so long as bodies are mastered and society stands in retrograde. Marxism believed that human equity would come through the domination of nature and in this 'it already displays the technocratic features later encountered in Fascism' to Benjamin's mind (393). Rather Benjamin looked to expose the destructive tendencies inherent in Marxism that made way for the ascension of Stalin. In the 1930s and 1940s, the Soviet Union violently purged many of its leading ecological thinkers and gravely degraded its environment in the quest for rapid industrial expansion. In the West, the Soviet system is believed to have faltered along technological lines to achieve its ambition. Its resurgence today in the form of the Russian Federation as a political force lies perhaps not surprisingly with its superior digital capabilities - in particular in the areas of intelligence and counterintelligence. Its twenty-first-century digital mastery over the global distribution of disinformation, in particular, has provided it with a new means with which to regain credible sovereignty on the world stage.

In his documentary "Hypernormalisation" the filmmaker Adam Curtis examines how much of the news broadcast to consumers of mainstream media is hopeless, depressing, and above all, confusing to its audiences. The response to a stream of seemingly endless crises is an increasing sense of nihilism and defeat among its audiences. Curtis suggests that their response is not an incidental but rather a fundamental component of a relatively new system of political control that has its beginnings with the intellectual relativism of deconstruction that coincides with the decline of Keynesian economics and the maturation of neoliberalism in the 1980s. The underlying aim of such a system was to use conflict to create a constant state of destabilised perception. In order for this to be accomplished media outlets had to constantly broadcast events which kept happening which seemed inexplicable and out of control. Today Vladimir Putin, Donald Trump, Brexit, the war in Syria, migrant crises, random terrorist attacks carried out by white nationalists and Islamic fundamentalists alike, all exist as the long term effects of that system's applied logic.

Throughout these events, those who are supposed to be in power were portrayed as paralysed in their responses to them. Curtis argues this because no one seems to have a vision of an alternative future beyond the principle of greater and ever more sophisticated forms of managerialism. This dire situation, in terms of a crisis in political authority, emerged at a time when the boundaries between what was real and fake were progressively blurred through a variety of competing new media formats. As those formats advanced through the decades of the 1990s, 2000s and 2010s it became possible for Donald Trump to craft a telegenic image of himself as America's archetypical capitalist. His business was conducted within a system of reality based on the merger of corporate infrastructure and multimedia spectacle that ultimately would support his commercial credibility as a qualified leader of the 'free world;' a boss of 'reality' itself. Trump promised his followers a return to a simpler world, the possibility of a presidency 'without politics' in which the free exercise of market forces would resolve all political issues in the most democratic way possible, i.e. through the aggregate of people's commercial choices. Trump's presidency speaks so well to Brexit and other unaccountable political phenomena because they all rely on what Henry Kissinger called 'constructive ambiguity' – in other words, politicians misleading their constituencies.

Over the past thirty-five years a compliant media world has furnished the conditions for a profound cognitive dissonance to set in in the public imagination through the media narratives broadcast to them that which suggested they lived in a hugely successful system generating unprecedented amounts of wealth that was 'trickling down' to everyone, bringing about a general increase in living standards and overall happiness. This positive spin on reality was at odds with what the majority of individuals had in actuality experienced as neoliberal economics progressively took root. What the vast majority had experienced since the mid-seventies were longer working hours, increasingly precarious working conditions, stagnating wages, a dysfunctional housing market and the decimation of the welfare state. Unable to account for these losses they comforted themselves by believing they had "Friends," could become an "Apprentice" to great wealth, or spend their time not "Keeping Up" with the Joneses, but rather "with the Kardashians".

A cult of market positivity was created to support them in times of need, as a means of maintaining the pretence of a functioning society even as most of them sensed its institutions were rapidly disintegrating. Those in the media tasked with narrating the realities of the world no longer assumed their job was to expose lies and assert the truth, but rather to maintain economic, social and political stability. The task was not about changing or imagining the world differently, but rather to focus narrowly on self-expression and individualistic action as a means of enacting progress. Dissatisfaction or dissention away from the status quo was carefully channelled into social media outlets which convinced users that were having their say in cyberspace. However, their posts had almost no effect whatsoever in reality. These acts of political expression had no changing effect because algorithms ensured these posts were only seen by people who agreed with them, making of political opinion a kind of echo chamber mirrored in the 24 hour ideologically branded news channels; all of which profited from riding the waves of mass public anger to their exhaustive breaking point. This is why politicised actions didn't appear to change anything any longer, and explains how this is intimately related to the rise of a class of political technologists that are able to control and distribute media content and manipulate the electorate on a vast scale.

As this new century matures, polarisation has progressively become the conduit through which we perceive the political communication directed towards us. At the same time, we have become accustomed to a world where the media is increasingly the means with which our bodies are made subject to (pre)occupation and distantiated violence. Digital technologies and new forms of networked media are transforming the ways in which publics perceive threats to their survival. A pervasive atmosphere of data surveillance, analytical manipulation, and the aggressive extraction of personal information has come to abide reality and by one means or another are strategies of societal containment waged daily by governmental and non-governmental actors alike. The ubiquity of smartphones, internet access, gaming platforms and social media now function as transports for disagreement, evading the reality of violence and effacing the complexity of conflict, while at the same time installing fear and the rejection of alterity into our interior lives. These technologies ostensibly assure us greater freedoms and unlimited access to information, but at the significant cost of betraying the moral distinctions between battlefront and home front, between soldier and civilian, between war and entertainment, and between public and private terrains which have now profoundly blurred from the perspective of the informatisation of the contemporary consciousness.

Application

Digital consumers are being comprehensively trained to feel most at home dwelling in our interior realities, occupying themselves by striving to reflect the tech industry's restrictive managerial discourses of creativity and merit, largely without a sense of how through such values they are subtly being compelled to act towards their own containment. Meanwhile, the likes of Google, Apple and Facebook augurs a future of totalising computation as the cost extracted from their user's signing of a social contract guaranteeing their self-realisation. This mechanism of reactive human development functions as a closed loop, ultimately bent towards fitting the entire world inside one curving arc of radical insularity, where the functions of ordinary life are seamlessly mediated and modulated by this small group of corporate entities. Essentially the era of technological construction promotes the regularization and management of human behaviour in service to the immaterial organisation and distribution of monetised information.

In the early 1980s, Jobs' distributed free Apple computers to schools as a way to encourage very young children to become habituated to a computerised learning environment. This meant essentially training them to adopt the conditions associated with an emerging digital culture centred around the language of computation (Parikka 34). Jobs from the beginning of his enterprise is not so much concerned with the development of Apple's hardware, but rather its software. In contrast to hardware, software has the ability to psychologically and technically drill into the minds of its users. It has the capacity to train them to adopt certain techniques and habits of use that perpetuate a desire to consume Apple's line of products. This task was accomplished 'by tapping into what could be called 'the involuntary part of the human being' (34).

What Jobs involuntary training program for young children essentially did was to introduce cognition into the equation of capitalistic extraction. This took place initially within the space of educational institutions and was eventually transitioned into the space of the tech industries corporate campuses. Jobs' approach to his enterprise wasn't about the acquisition of technologies, so much as the establishment of a desirable interface with them. His unique approach would fuse previously known business strategies with emergent psychological techniques, compelling digital culture to conform to his hegemonic vision of the future. The future itself in this scenario could be prefabricated as a 'software,' housed within a territory of metaprogramming that would radically recondition labour itself, as both divorced from location and equally, boundary. Jobs' creative passion was, therefore, essentially directed at programming the programmers, so that the brain itself was mobilised in the cause of generating capital via the monetisation of information.

Jobs modelled his postmodern ambition on the computers that worked before and during World War II. During this era, computers were about teams of humans, predominantly white and black females, managed as computational units. 'The organisation and management of humans labouring as machines was what defined the early "computer architecture"' (Parikka, 45). Jobs' first piece of software devoted to this ambition is the Lisadraw software, which allows the user to manipulate visual objects on the screen of a computer. The techniques and abilities this software fosters and redirects notions of artistry and creativity in post-Fordist economies and software culture, making inventive productivity synonymous with the skilled manipulation of computational units. The Lisa of Lisadraw, Jobs' own daughter is atomized from a previous architecture of pluralistic problem-solving. The labour implied in her programmer name, 'Lisadraw' remains coded as female. Nonetheless, the Lisa behind Lisadraw is notable in her absence from Jobs' original demonstration of the software. It is he, not her, who manipulates this new rendering technology to 'move, to shrink, change textures, airbrush, hardened and soften edges – all of which makes him realise how the talentless can draw' (Parikka 34). The pleasure of this exercise derives from the fact that one no longer has to rely on the talents of others to succeed. One no longer needed to have the talent to draw in order to create. What replaces drawing as an analogue age talent, is the digital repositioning of creativity as a modality of self-expression.

Silicon Valley's strategy to program the programmers allowed those privileged enough to play, relax, experience, and innovate within their meticulously designed platforms to do so without ever encountering those involved in the material labour that maintains their curbed virtual habitation. The idealisation of the commons inherent in platforms, such as Facebook, Twitter, Youtube, and Instagram at base promote concerns having not to do with reality, but rather reality management. These channels cast themselves as inclusive, aspirational, and ideational. In this way, users are sheltered from the blunt appearance of any material obstacles that might lay beyond them. In such instances, individual creativity becomes a substance born of subtle, prefabricated direction. On these platforms, the user becomes trapped in a never-ending quest for continued inclusion, lest one should 'fall from the ranks', and find themselves amongst 'the surplus populations' that failed to gain and maintain recognition here, and as a consequence 'are rendered invisible, voiceless, and ultimately non-existent' (Pinto and Franke 29).

Within such a terrain the digital promotes an imaginary image of itself as a conduit of social emancipation. In reality, it tethers its user to a future of tribalised identifications. Dating back to the Whole Earth Catalog, cyberspace was initially conceived of as a meritocracy that would will naturally and inevitably build virtual communities based around the shared values of creativity, achievement, inclusion, equity, resilience, and sustainability. Nevertheless, it remained the case, that those that went on to be 'makers' of these spaces were disproportionately young, white, and male. These individuals generally didn't worry about the military-industrial community that was already there per se, but rather saw the potential to appropriate a space within that for the privileged few who already had the luxuries of time, wealth, and technological literacy at their disposal. The arrival of the Internet puts into motion a sequence of events that will ultimately drive them into positions of authority.

By the mid-1980s, both corporations and their employees find themselves dwelling within a new era of capitalism where competitive processes accelerated through the algorithmic transformation of market relations. At this time, there was a sudden deregulation of financial markets, which coincided with the introduction of electronic, screen-based trading. It was referred to at the time as the Big Bang as it marked the beginning of a worldwide financialisation of capital. As a consequence, those individuals working across finance, venture capital and technological innovation found themselves part of a burgeoning cosmopolitan global elite. Moreover, they found their speculative interests now increasingly converging with those of the emergent middle classes, political elites, and university-based young entrepreneurs. This is no longer a question of supply and demand of capable people, but rather of exploiting the minds of these constituencies to believe that their power laid ultimately in their ability to assimilate and produce knowledge. Within this enclave, information was used as part of a strategy to influence subjectivities and behaviours in order to make them more congruent with the neoliberal principles of the economy. As such what is being witnessed in Silicon Valley should be perceived as a demonstration of the logic of the contemporary neoliberal restructuring of the mind to coincide with corresponding digital geographies that generate new behavioural structures and psychosocial processes.

At the centre of this was an expanding digital infrastructure that was constantly busying itself with the ranking of individuals according to specific abilities valued by the technology sector. Therein users were subtly compelled to adopt social behaviours and patterns of consumption that complement the trend towards digitisation at the same time that Silicon Valley positions itself as one very large incubator for profitable innovation. Those able to optimally produce content within these systems were more often than not well-resourced men who were not expected to engage in social reproduction. Silicon Valley's corresponding geographies marked out 'an ever-expanding sociospatial terrain of governance – across digital and physical space –where conjuring creativity and innovation is an omnipresent objective' (Sanson 11). In these instances, the community in question was increasingly coerced to become resident to labour in the relentless modulation and socialisation of what might be considered "workspaces." 'Such processes reorganize our lives, rather than our labor, around the pursuit of capital, and value extracted by creative workers is dispersed across individual firms, private property developments, and local and national cultural enterprise strategies' (Sanson 14).

This model suggests that computing should come into all aspects of what we perceive as lived reality. It assumes that the independent decision making of a great number of individuals coalesces into a kind of unconscious intelligence about the environment and its broader relation to a political economy, based on an accelerated response to difference, assimilated into the capitalisation of interactivity as a whole. In this context, cognitive labour becomes synonymous with personal enhancement and lifestyle management. The emphasis here is one that promotes focus, well-being, and pleasurable interactivity with both like and likeminded others. At a deeper level still, it perpetuates a symbolic economy of cognitive influence, affective capital, continuous productivity, and decentralised hierarchy. These corporations have learned over time that value can be extracted from monitoring the living reality of communication and affective social life within the creative, semi-autonomous ecosystems it generates to expand capital and define value.

The End of the Future begins under the auspices of liberalism the wherein geosphere was largely elaborated through spatial analogies that foreground a physically tangible division between the public and the private, between rights of ownership, control and usage. The neoliberal geosphere of data can no longer be comprehended in such terms – as a static, albeit progressively opaque, environment that we simply traverse. Rather, today's global imaginings bring with them a profound blurring of the capacities, roles and motivations of different actors. Traditional power apparatuses are now being confronted with populations that are increasingly tasked with the self-servicing of the social, cultural and infrastructural fabric of the societies they inhabit. Within these new spatial coordinates, any notion of social capital is now joined to an affective economy powered by desire, identification, apprehension, and capitulation. The fate of these new data publics is thus both generated and implicated in new technologies of governance and their associated digital economies that presently cross into and beyond the realm of data. The acceleration of data constitutes one of the most powerful transformative forces in the world today and is radically changing both the way we live our daily lives and processes operating on a global scale. This book encourages an active understanding of the implications of hybridised data environments in shaping individual, commercial and governmental agendas and actions. Through both theoretical and historical explorations, it enquires into the impact of a 'data mentality' on our expectations and articulations of the future. As governments become increasingly private and citizens increasingly public, it is vital to consider what knowledge might be utilised to expand an awareness of data as a provenance accepting of both new forms of public order and private domain.

Detonation

This book commences its journey with its first chapter "Conspiring Worlds: Emergent Sovereign Imaginaries in the First and Second Late Elizabethan Age". This chapter takes as its starting point a humanity that now finds itself captive to pervasive institutionalised forms of violence whose force has everything to do with the aggressive pattern of economics in a neoliberal age. Humanity is now thoroughly captivated by an Internet that has taken on a fourth dimension to generate consequence in the material world. This has allowed the Internet to become thingly insofar as it will soon be understood not as an interface, but as an environment. As such it has taken on the ability to shape conditions beyond the imaginary and embed itself into materiality in a variety of ways that benefit from the demise of state parameters and the enlivenment of a fluidity of information able to migrate across time and space.

Elizabethan England was an era of unprecedented speed when it came to both navigation and information. Hers was an age of data, wherein traveller's reports could circulate in tandem with the geographical inscription of new lands onto a virtual mapped reality of an England to come, through the means of nothing greater than information itself. Coupled with speed, communication bore a path to understanding that fascinated philosophers, military tacticians, and magicians alike whose formulas and equations acted as progenitors for thinking a universal wisdom, an 'online' language that could be imprinted on the world as logically as longitude and latitudes had. This language had to be artificial in nature, an encoded language that outpaced the need for translation and allowed an instant retrieval of higher forms of perception.

We have now entered into a phase of existence where information now travels faster than people or things, this is what McKenzie Wark terms 'third nature', which summons into being a space within which all the possibilities for the organization of other spaces come together. As this phase reveals itself, what the general populace takes to be a natural, or naturalised set of circumstances on which their day-to-day existence is dependent, is what Wark refers to as the 'peculiar geography' of a thoroughly mediated, vectorialized world. Trauma ensues in the instant the human world assumes recognition of itself as data and its basis for existence, one fully subordinated to the software processes that tabulate, index and sort the entailed relations making up the human world. The acceleration of data threatens to relinquish ephemeral modes of representation to ceaseless processes of computation.

All of this is significant insofar as it relates to the advent of what Simon Critchley and Jamieson Webster term 'a second late Elizabethan age,' which in essence means that within contemporary England policies of militarization and surveillance are paramount to the conduct of ordinary life; one that is at once supported and contested within the scope of contemporary digital media. Within this highly mediated macrocosm, it must be assumed that everyone is being watched and moreover, that everyone lives in fear because they know they are being watched. At the same time, they exist in a state of distraction, as they try to live amidst a perpetual crisis of legitimacy installed to occupy their thinking as a complement to the paranoid reality of political life. Within such a schema humanity's relation to the future is not one of being and nonbeing, but rather an eventuality that can be mapped on a continuum extending between absolute control and total submission. With some futures considered more viable than others, humanity's ongoing capacity rests on its powers of discernment amid a vast array of seemingly heterogeneous possibilities that seldom find their resort in the assignment of material value. A new geography emerges based on computation to centralise the meaning for the subject, while at the same time burdening it with a new algorithmically mediated self. Diverted away from its former material and organic capacities, it now finds itself diagrammed into a new protocol of infinite connectivity.

Chapter 2 "Poverty and Experience: Walter Benjamin's War in the Age of Vertical Sovereignty", critically re-examines Walter Benjamin's canonical essay "The Work of Art in the Age of Its Reproducibility" to glean its critical portent for our current age of digital technology and remote warfare. The particular issues we are wrestling with today in regard to new imaging technologies are derived in part from the very systems and techniques of domination that Benjamin began to critique over a century ago. As we confront a critical terrain littered with concerns about neoliberal governance, neocolonial warfare, environmental denigration, the global preponderance of new forms of social media, the deepening of social and economic barbarity, and an increasingly violent political landscape, we must remind ourselves that what we are witnessing today is an intensification of the broad global trends that Benjamin sought to identify and understand. Benjamin's montage reading of history takes place amidst a crisis in democracy, wherein mass media has triumphed by transforming politics into a screen and history into a theatre of war mediated by technology. Filmic imagery functions as a temporal index that enables fragments and moments of the past to intrude upon present experience. Such images preserve experiences, which can be re-enlivened and redeployed elsewhere in history in aid of the possibility of colliding with their present and future viewers. In the twenty-first century, that screen has split off from itself and proliferated onto multiple surfaces. As a result, contemporary political power lies in the control not of the means of production but of information. Thus, it is possible for an inverted image of totalitarianism to flash up at this point in history, and be altered to support our present condition of binary apperception. Digital media now serves to legitimise and promote a customised micropolitics of identity management, in which each human soul is captured and reproduced as an autonomous individual bearing affects and identities. The digital infrastructure of the twenty-first century makes it possible for power to operate through a mathematical means, and for factual material to be manipulated in the interest of multiplying the means of control.

Chapter 3 "Black Mirror: Binary Apperception and Digital Racism", explores how the rapid proliferation of access to digital platforms and social media sites has radically refigured the terms and topography of racial representation, politics, and cultural expression. Within such domains race is being subtracted from itself on a microlevel and lent a verisimilitude of essential purity in which certain racial subjects are portrayed as dominant, and all others consigned to inveterate alterity. Digital media now serves to legitimise and promote a customised micropolitics of identity management. Ethnic and racial coding has become synonymous with mediation itself and the fate of bodies determined by the way the digital apparatus compels them to perform restively within a certain affective identity's limited contour. The passive reception of what is being projected cannot be classified as a leisure activity, or a frivolous encounter so long as the masses continue to perform the productive activity of making operative a system of subjective predication. Digitisation has emerged as an apparatus for reinforcing this order, through its assignation of certain identifiable traits to distinct races within society and manufacture of complex narratives for explaining and maintaining such hierarchies as an intrinsic phenomenon.

Since the dawn of the Cold War, race has moved beyond the limits of a modern and liberal biological determinism to become a dynamic mode of transmission of codes and information, its former ontological position now opened outward to accommodate the mutability of new imaging technologies, made both differential and communicative in their design. In the current postsocialist neoliberal age, the screen has eclipsed the lens when it comes to the mediation of race as a feature of embodiment, its colonial origins largely erased from our contemporary understanding of it as a source of both truth and conflict. As such the hypermediated raciality of the past four decades must be approached critically, not as something that offers greater self-understanding, but rather something that gives a crucial insight into the ways in which humanity is being aggressively mobilised to reproduce a society of control through our fascination with the capture and differentiation of ourselves and others.

Chapter 4, "Big Blue: The Anthropocene and the Advent of Planetary Digital Warfare", starts from a premise that the end of the future can be said to have begun in the year 1945. This is the year our blue planet was reconfigured to obey the coordinates of an emerging technology culture whose geopolitical imprint specified a necessity for life on Earth to mobilise in response to what was deemed its "Great Acceleration", its forward thrust towards its own potential annihilation, from the bankrupting of its natural resources on one hand, and on the other, the fragmentation of life itself to the ontological status of building blocks and components. This would be the era where life and its artificial counterpart would begin the long course of their corporate merger in earnest. It is also when life would submit to a new radical manoeuvring, through a globally-orchestrated technic of politics and economics, bent on striking a balance between the emergence of planetary-scale technology and micro-scale technology through a variety of experiments related in one and another way to social engineering. A new sense of reality itself was effectuated by subtle manipulations of algorithmic code, and discipline of our imaginations through cybernetics, allowing for new epistemologies to emerge. Through them, humanity began to conceive of our Earth as a technosphere and our mission to mine its resources, in order to platform new worlds to service our material dependencies to come. With the advent of this new age steeped in fascination for the atomic, digital, molecular, and inorganic, humanity came to fully conceive of itself as the agent of its own operational destruction.

The activation of this pattern of thought permitted a type of humanistic liberation theology to come to the fore that swept its way through not only the intellectual domains of research and development, but equally philosophy and aesthetics all of whom were eager to take their place as actors in what was most self-consciously deemed a revolutionary era for humankind. And so, it was that we entered unwittingly into what we now commonly refer to as "the Anthropocene". We now readily engage with the performative elements of this terminology whose advent neatly coincides with the historical invention of "the environment" as Earth's global operating system, the elevation of art and culture as mediums of calculated experimentation, and the integration of Silicon Valley as the global nexus of a rapidly ascendant military-academic-industrial-complex. We dwell now on a planet littered with the remains of these largely destructive postwar conceits, taking their form variously through the appearance of ubiquitous petrochemical waste, the wholesale corrupting of nature through deregulated international trade and foreign investment, the advent of new forms of race, class, and religious disaggregation, the monetisation of information, and the mass replication and storage of data. How can we resist this recursive stance towards the universal back-dated now to the post-WWII period, and at the same time achieve distance from the tyranny of contemporary neoliberal commensurability? What forms can be sought beyond the optimisation of reality as an infrastructure for controlling human subjectivity that might reveal in its place the asymmetrical articulations of sovereignty, the impositions of a techno-scientific elite, and the structural violence of institutional protocols throughout the globe that subtends such a dire conditioning?

Chapter 5 "Coda: The Necessity of Reattunement", picks up from Chapter 4, to explore the connection between the popular culture of the 1960s and the counterculture. Exploring the music of the Beach Boys, The Beatles and Sun Ra, it addresses sound as an instrument of oppositional consciousness, particularly in relation to struggles over space. Cultural phenomenon unique to that period including the Apollo space missions, Charles Manson, the Black Panthers, the Beatles "White Album" unite in the cause of seeking salvation and renewal amidst a backdrop of social chaos. Each in their own way emerges as a key formation related to the counterculture in its quest to join the satiation of personal desires to a greater project of expanding social freedoms. The libertarian values of 1960s counterculture celebrated a lack of personal responsibility in light of a larger goal. This value system was informed by a heady mix of sexual promiscuity and excessive drug use, weak personal ties, casual interpersonal relationships, and a strong yearning to interrogate and disrupt societal norms. Popular music was seen as a conduit to promote these values. At the same time, vibration held greater political connotation as a means of extending understanding outwards. Through sound broadcast society found revolutionary ways to promote and sustain deep political introspection. This probing yielded both great violence and a greater cultural reckoning. Walter Benjamin asserted, that what characterises revolutionary classes at their moment of action is the awareness that they are about to make the continuum of history explode allowing for another kind of consciousness to come into its own.

This book ends with the "Epilogue: The Futures of Ends" which allows us to consider how the first golden age of corporate privateering of wealth and plundering of bodies, the late Elizabethan Age, has grafted itself onto a future where sophisticated technologies of surveillance, carefully mapped raids of homes, systematic torture and state-sanctioned murder against those identified as enemies of that state would go onto establish normativity across some five millennia. Progress in these directions was meant to grant society the gift of stability and prosperity, the cost of it remaining obscured for those in the highly lauded position of beneficiary. Little is now said of how civilisation was accomplished through these techniques of power and how information technology at every stage furnished their arrival. Both time and energy have been commandeered into this cause for which appropriation and co-optation of the property of others seemingly have no limit, born as were from the insatiable drive to master fortune, territory, and intelligence. The vast consequence of that endeavour has forged the very substance of our contemporary capitalist universality. A minority seeks to question what lies beyond the ever-receding promise of the optimisation of reality. Nevertheless, it is only in seeing through inveterate progress of technology that an alternative means of infrastructure and of patterning subjectivity can be sensed elsewhere and otherwise. The trick to humanity's survival might well be to allow its proscribed ending to come without further ado, motivated, as it were, by the unsteady reassurance that from even the grimmest of endings springs forth the prospect of indeterminate new beginnings.

Chapter 1
Conspiring Worlds:
Emergent Sovereign Imaginaries in
the First and Second Late Elizabethan Age

'Information wants to be free but is everywhere in chains.' - McKenzie Wark, *The Vectoralist Class*

'The much-vaunted "end of history" may be an ideological phantasm, but there is such a thing as the end of the future—or, at least, a crisis of futurity.', - Hito Steyerl, *Postcinematic Essays After the Future*

'We have no future because our present is too volatile. The only possibility that remains is the management of risk. The spinning top of the scenarios of the present moment.' - William Gibson*, Pattern Recognition*

'Empire does not confront us as a subject, facing us, but like an environment that is hostile to us'- Tiqqun Collective, *Introduction to Civil War*

In the twenty-first century, the future has become a universal property based solely on information. What had been of material value for much of the twentieth century, began in in its final decades, to transfer its worth into the realm of raw data, forcing the self to reside in a place of constant and perpetual exposure, and its surrounding community to linger faintly amongst the remains of its former desires. We have become content to simulate the future rather than generate it. This has been the case since the early 1970s when information technologies began their popular ascent. These technologies accelerated us into an era we referred to in a pre-millenary moment as 'postmodernity'. Its hallmark was to postdate reality and replace it with what 'Jean Baudrillard and Eco used to call the "hyperreal" -the ability to make the imitation more realistic than the original' (Graeber 110). The postmodern prompted us to delight in the fact there was nothing new to look forward to and therefore our sensibilities must rely on the clever commands of 'simulation, ironic repetition, fragmentation and pastiche' (110).

Baudrillard characterised 'hyperreality' as a form of deterrence with its roots in capital accumulation whose historical trajectory was bent toward 'the destruction of every referential, of every human goal, which shattered every ideal distinction between true and false, good and evil, in order to establish a radical law of equivalence and exchange' ("Simulacra and Simulations" 182). He argued that capital was the first technology 'to practice deterrence, abstraction, disconnection, deterritorialization' and moreover, that throughout the intervening centuries of its mature as a power, it 'was capital which fostered reality', and 'the reality principle' (182). Beyond this, what was crucial to recognise for Baudrillard, was that capital was also 'the first to liquidate' reality 'in the extermination of every use value, of every real equivalence, of production and wealth, in the very sensation we have of the unreality of the stakes and the omnipotence of manipulation' (182). Baudrillard's description of the logic of capital is as a kind of privation of reality, which eventually leads to the appearance of its catastrophic remainder, as it travels at warp speed into the sudden deregulation of financial markets and into the new dimension of electronic, screen-based trading. If capital was always already a futural entity, its unprecedented acceleration in the early 1980s, signalled that reality - as we had come to know it - was in a death spiral, and what would come to replace it was a multiplicity of signs and a theatre of simulation.

The global convergence of capital and electronics meant that the future we dreamt about previously would never appear in our outward lives. Rather its contents could be said to have imploded, its flattened image compressed down to nothing more ambitious than what can be fit within the ultrathin surface of a screen. In the race towards the future, it would seem the computers bested our imaginations. Ours was to become a technologically mediated environment and creation itself recast as a process of transfer of information and rearrangement of control played out amongst existing forces - albeit somewhat modified ones. In the twenty-first century, the means of production are no longer functioning as a locus of power; that power now resides with the ability to discipline the velocity and magnitude of information. The objective of its evolving class of leaders is not to seek rent, nor profit, but rather interest.

McKenzie Wark refers to this new interest class, as 'the vectoralist class' and argues that its emergence has only been possible through 'the production of an infrastructure in which information becomes separated from its material strata', allowing for it to be 'efficiently channeled anywhere on the planet, stored at negligible cost, processed easily into complex patterns' ("A Haker Manifesto" 5) This new infrastructure initially functioned as a complement to 'other kinds of control through information, whether through the control of patents, copyrights, and brands, the control of supply chains through logistics, the control of the spatial deployment of resources through GIS' (5). Through such mechanisms the primitive accumulation of a previous capitalist modernism has been overtaken by interest accumulation, which seeks something greater than merely a return on investment. Rather its wealth is derived from a surplus of information. Its project is no less skewed than its former manifestation. What had been previously acquired through unequal exchanges of global capital, now tips the balance through global information. Europe and the United States still dominate in this arena spawning in its wake, a vectoral class which thrives by extracting surplus information on a global scale, and in the process, transforms capitalism itself from a postindustrial concern to a managerial function.

No longer dealing with the more routine means of production, this new form of capitalism can now focus its energies on barring access and controlling information. The effect of all of this is that we now reside in a dominant infrastructure increasingly designed to the specifications of the vectoral class, which subordinates our concept of reality to conform with the expectations of hedge fund managers and CEOs projected outward to all classes below them through the subtle tyranny of screen technologies. We are no longer directed to the spectacle of the major breakthrough, but rather towards a fascination with modest tweaks to existing organisational structures. We are no longer curiosity driven, but rather risk adverse societies. We occupy, not an age of discovery, but rather one of retrenchment. Austerity dominates our current thinking, and forces our concept of technology to become limited to a means of arriving at ever more sophisticated forms of information and imagery, rather than advancing a better tomorrow.

Historical Hacks

In the summer of 1989, Francis Fukuyama declared that the end of history would become apparent through 'an unabashed victory of economic and political liberalism,' wherein 'the total exhaustion of viable systematic alternatives to Western liberalism' could finally and universally be embraced 'as the final form of government' ("The End of History?"). Politics as a technology of the future in this crucial moment had been conquered, as evidenced not by its domination of 'the real or material world,' but rather 'the realm of ideas or consciousness'. This is crucial because victory could be claimed as a consequence of bombarding developing nations, with consumerism, rather than securing their ideological acquiescence to Western liberal democracy. Fukuyama's assumption was the mediated ideals that are projected upon them over time would eventually come to govern their reality. Within such a scenario the commercial would precede the political as a mode of universal homogeneity and would arrive at the behest of easy access to VCRs and stereos' as both 'a symbol and an underpinning' of democracy itself.

Fukuyama predicted that 'the end of history will be a very sad time' as it would be marked by the triumph of the technocrats over the revolutionaries ("The End of History?"). Fukuyama lamented, 'The struggle for recognition, the willingness to risk one's life for a purely abstract goal, the worldwide ideological struggle that called forth daring, courage, imagination, and idealism, will be replaced by economic calculation, the endless solving of technical problems, environmental concerns, and the satisfaction of sophisticated consumer demands.' Dissent would be eliminated in this scenario of the future, and spirit vanquished by a complex managerialism. Perhaps the most disturbing of Fukuyama's assessments was that in 'the post historical period there will be neither art nor philosophy, just the perpetual care taking of the museum of human history' and a preoccupation with nostalgia regarding our dynamic past. The utter boredom associated with these tasks, Fukuyama hoped, would eventually motivate humanity to revive history itself.

Reckoning with the Postdigital

There is evidence that our enchantment with the digital age is waning and that humanity has begun to pilfer the museum of history, as a means of countervailing the forces of technology hegemony. Even as our contemporary consciousness becomes progressively more pervaded upon by a plethora of mobile devices, the commercial expansion of drones outside of the military context and the ubiquitous data collections of agencies like the NSA and GCHQ, and the monetisation of our identities and preferences through Google and other global social media platforms the revival of interest in analogue devices of the past allow us to experiment with a deliberate repudiation of this reality.

There is an increasing desire to emerge from the dystopia of post-history where the smartphone relay of terrorist attacks, mass shootings, or other forms of extreme violence such as beheadings, bombings and violent murders perpetually disrupts consciousness and mollifies thinking. The period in which our fascination with these systems and gadgets may itself be becoming historical, as the physical reasserts its value within the two recent generations, Y and Z. The digital had commanded their constant attention, subjecting them to multi-screened and multi-tasked existences, that reinforced their acceptance of a world steeped in perpetual struggle, inequality and brutality, where their desire to have agency and impact would only be temporarily realised under fleeting circumstances. Their soaring interest in typewriters, vinyl records, audio cassettes, as well as analogue photography and artists' printmaking, speaks to their desire to resurrect the potential of an alternative future manufactured through these pre-digital devices.

Their recourse to these older forms of media is not intended to sever a connection with digital communication, but rather to pragmatically and meaningfully augment popular, cultural, and colloquial frames of reference. By installing within them a fault aesthetic that conspicuously repudiates the objectivity and sterility associated with advanced digitisation, they are able to dynamically reject 'the "idea of digital progress" as well as "a teleological movement toward 'perfect' representation"' (Andrews qtd. in Cramer 20). As Florian Cramer argues, this approach also counters Fukuyama's hypothesis, insofar as the concept of the 'post-digital' was conceived 'primarily as an antidote to techno-Hegelianism' through on one hand a renunciation of innovation, improvement, and higher fidelity and on the other, a denunciation of the paradigm of 'technical quality altogether' (20).

The post-digital disorders universalism and the automation of information by reducing a primary means of production to a secondary means of reproduction. At one level, the post-digital stands in direct opposition to new media. On another, it function's allegorically as new media's 'negative mirror image', exposing the weakness of both its historico-philosophical connotation, and teleological foundation (Cramer 15). The dubiousness of the pre- fix 'post' degrades the categories it wishes to define and in so doing corrupt the viability of the techno-positive narratives that labour to reinforce their connection. The post-digital provides younger generations with a means with which to reject the self-affirmative discourse of immateriality promoted by Silicon Valley futurists and recommission the materiality of race, sex, and class status that consequentially determine their future.

For them these formations are not cultural artefacts to be relegated to the dustbin of history under the proviso of a neoliberal meritocracy, but the codes and texts that form the bases of their conditional exploitation. The residues and circumstances of their categorical making are what concern them most in terms of anticipating the limitations assigned to their situation. Unlike their generational predecessors, their perception of society arrives through a mastery of its imperfections and malfunctions. As such they are able to readily assume a hacker mind-set of 'taking systems apart and using them in ways that subvert the original intention of the design' (Cramer 20). Theirs is a guerrilla positionality attuned with other postcolonial practices that seek to resist the prescribed patterns of understanding enforced by 'what Thomas Elsaesser calls "military-surveillance-entertainment complex"' made up of only a handful of global players intent on controlling the form and movement of information (Steyerl 14).

Like any good resistance movement, it must mimic the rhetoric of its master. In this instance, a hegemonic cadre start-up and global technology companies powerfully mobilised through the language of freedom and individual empowerment and carefully sited through a fiction of self-making. 'Freedom with and of information is the utopia of the hacker class' and therein lies the danger of balancing resistance with a desire for agency as we enter in a new era of 'maximum control coupled with intense conformism' (Wark "The Vectoralist Class" 20, Steyerl 17). As enterprise becomes freer individuals become progressively more enslaved by 'the hard and fast automatisms are introduced into material relations between people' (Berardi 105). As the cultural and juridical conventions of modernity and bourgeois law begin to rapidly degrade a new hegemonic force emerges rendering all bodies universally subject to a non-consensual technological restructuring.

Dehumanising Labour

Humanity now finds itself captive to pervasive institutionalised forms of violence whose force has everything to do with the aggressive pattern of economics in a neoliberal age, bound with an internet that has taken on a fourth dimension to generate consequence in the material world. This has allowed the Internet to become thingly insofar as it will soon be understood not as an 'interface but as an environment' (Steyerl 17). It thus takes on the ability to shape conditions beyond the imaginary and embed itself into materiality in a variety of ways that benefit from the demise of state parameters and the enlivenment of a fluidity of information able to migrate across time and space. All previous forms of media suffered from imprisonment within a screen, which limited their ability to function as the foundation for alternative networks, or as nodes of multilayered connectivity. The Internet of Things summons forth nothing less than a new form of imperialism able to transform space into a sphere of liquidity, and complexity into a condition of movement. As labour precarity and labour migration becomes the normative situation of a disenfranchised humanity, so too does its acquiescence to a universal accessibility where they and their belongings are continuously subject to digital interpolation, and as such their behaviours and movements made available to generating exploitable forms of interest. The data we create is thus credited to others and interpreted to advance interests of others; all in order to bankroll the states and the corporations who control the means for 'making worlds we alone discover' (Wark, "A Hacker Manifesto" 2).

Similar to prior forms of imperialism, control is built into the very design of the infrastructure that allows the vectorialist class to dominate those embedded with their systems of operation by reconfiguring class strata to conform to new standards of ownership. Meaning essentially, as McKenzie Wark acknowledges, that in the neoliberal age of monetised information 'we do not own what we produce - it owns us' (Wark, "A Hacker Manifesto" 2). "Bifo" Berardi contends that 'the dynamics of neoliberalism have destroyed the bourgeoisie and replaced it with two distinct and opposing classes: the cognitariat on the one hand, i.e., the precarious and cellularised labor of intelligence, and the managerial class on the other, whose only competence is in competitiveness' (104). The space of contestation between them 'becomes a global topology in which almost any point can connect to any other, mobilizing resources on a planetary scale' (Wark, "The Vectoralist Class" 8). This generates a crisis within an economics modelled on a prior division of labour, between thinking and doing and transfers its uncertainty outwards towards the development of a post-historical capitalism, which in Fukuyama's estimation would 'be frictionless, immaterial, labour-free' (Mcclanahan 87).

Within this model labour's apparent disappearance was in reality going to be a dislocation. As productive labour moved to the global south casualised, precarious forms of labour rooted in the global north, primarily in administration and the service industries. If the abundant end of history brought unprecedented debt and economic insecurity into American and British households, in the developing world things were far worse. The production of post-industrial technologies operated through a return to a productivity pattern reminiscent of nineteenth-century factory work, replete with child labour and the extension of the working day. These 'historical forms of domination and power, encompassing but not limited to social categories and hierarchies of difference' were retooled for the information age and as a consequence of the desire for absolute extraction of surplus value, subsequently 'get built into seemingly non-human objects' (Atanasoski and Vora). The racialised and gendered human labour that underwrites digital media production is obscured by the corporate infrastructures that control their movement across markets and in the process, sanitise them to appear as human-free entities and thus play a crucial part in shaping our perception of who, or what can count as human. The ultimate goal of their continuous development is to produce a surrogate humanity who performs degraded work that is always already meant to be invisible.

Invisibility had arrived even as digital servitude had not yet been fully achieved beyond the current biological-economic imperative. This imperative developed alongside and through the demands of colonialism, remains crucial insofar as it designates 'a distinction among human beings ... such that humanity was something to be achieved' (Scott qtd. in Atanasoski and Vora). Civilisation, therefore, has long been synonymous with European technological innovation as progress, itself pivoted on the ambition to humanise that which is Other. An ongoing differential achievement in the status of the human serves to heighten competition and a redistribution of wealth gained through neoimperial exploitation. The introduction of posthuman splinters a previous capitalist imperial paradigm based on biological race, and reconfigures value and valuelessness through new paradigms of division associated with biometrics and information.

As a coming race, posthumans enable their masters to conquer the last bastion of outsourced labour as these bioengineered machines will now do not only the making, but the thinking for us achieving slavery's ultimate goal which was to produce a sub-humanity that was infinitely reproducible and consumable at virtually no cost beyond the initial investment. This technological innovation can bring about such an elimination of surplus workers altogether in favour of the employment of cheaper, efficient, and productive information technology, robotics, and artificial intelligence violently reducing their value to merely that of function. The collapsing of value of the human implies the eventual removal of actual labouring human bodies to the background of life so that an elite can more freely occupy the foreground 'transposing categories of abjection from the human into the very infrastructure upon which the techno-utopic future depends' (Atanasoski and Vora). What is perhaps most valuable about the posthuman is not its ability to think for us, but rather its inability to think for itself. Unlike the labourer who precedes it, its mind is one that can never arrive at a political consciousness, or question the economic or political system within which it reasons. The posthuman is also critically incapable of forging its own imagination and lacks the creativity or desire to challenge power relations, or to fundamentally recognise its inferior positioning.

There can be no coincidence this comes at a time when neoliberal economics has its logical conclusion in a totally abject and submissive workforce beholden to an elite class. Similar to the liberalism that preceded it, neoliberalism depends on 'global divisions and asymmetries' through which certain liberties such 'as the universal promises of rights, emancipation, wage labor, and free trade' are 'reserved for some and wholly denied to others' (Lowe 3). Stark differentials in 'individualism, civility, mobility, and free enterprise' have always been at the heart of capitalist innovation bringing forth 'new means and forms of subjection, administration, and governance' that have their beginnings with the founding of new markets (3). This gross discrepancy in global authority quite literally is meant to go unseen by consumers who find themselves affectively conditioned to embrace the Internet as a productive space, which nurtures, sustains, and exhausts them into producing its essential value as a class liberating activity. The apparently free quality of 'liking' on a Facebook page, or 'wanting' on advertisements linked to browsing activities, comprises their desire for self-expression into unpaid manual labour as the capitalist economy's demand to service information becomes seemingly insatiable. What we assume is a brute binary of division between self and other in the neocolonial space of the Internet emerges as an operating system based precisely through spatialised and temporalised processes that both differentiate and connect.

In the labour-free, posthuman context of production it would not be unthinkable for various elements of society to be excluded, removed, captured and killed in the name of progress and capitalism. While it may be possible to keep the overt horrors of production within the global South out of view of most American and European consumers, its brutalities have nonetheless perverted and troubled its cultures, values, and tastes. At some level, they must concede that the functioning of their societies was only possible in relation to 'the colonised geographies or "zones of exception" with which they coexist, however disavowed' (Lowe 17). These same occupied territories make possible 'the management of life and death that we now associate with neoliberal security regimes and the state of exception in crisis and [perpetual] warfare' (Lowe 17).

Steady State Apparatus

Global economic and military expansion has been used as a means to manage the domestic population since the dawn of Britain's first era of colonialisation during the reign of Queen Elizabeth I. The colonies were never just a means of resource extraction, but equally served the function of a depositing ground for its surplus population who could, or would not be functionally employable at home. Commercial advantage gave lifeblood to the English version of global hegemony, which sought to map the world to the specification of the ruling elite redlining both its domestic and imperial policies to articulate a new England, where governance coincided with the advent of a surveillance state at home and military interventionism abroad. The Elizabethan Empire relied upon a potent combination of domestic terror, religious tyranny, foreign privateering, and covert espionage to secure its purposes. The mechanisations of empire often lead to violent death via political assassination for those oppositional to the ruling class, or religious murder for those subversive to Protestant hegemony. Loyalty was equally treacherous to England's commoners who suffered the dire effects of military recruitment due to the fact that many 'died not in battles but as a result of the hardships of service', more or less as the lower-class victims of an upper-class appetite for endless conflict (Shapiro 26). War-making thus effectively functioned as a means with which to dispose of commoners as one amongst the many abuses enacted through the Elizabethan security apparatus.

This was made manifest not solely through armed conflict, but also through a medieval modern state apparatus that aggressively 'manufacture[d] itself by destroying alternative conceptions and practices of power' (26). Curtis C. Breight argues that the sole objective of Elizabeth's regime led by William Cecil, Lord Burghley, was the fortification of domestic control and that techniques of surveillance and terror were developed expressly toward that end. This took place variously through 'extension of treason laws, creation of Provost Marshalls, expansion of judicial torture, invention of an espionage network, fabrication of pseudo-plots, staging of kangaroo-court treason trials, springing of mass house-to-house searches in London producing huge and detailed intelligence documents' or the 'sustained promotion of both propaganda and paranoia (executions reached 800 per year by the last years of the reign)' (Fitter para 2). All of these formations were united as 'manifestations of an aggressively expansionist state apparatus' (para 2). Elizabethan England was effectively a police state where multitudes were subject to what amounted to totalitarian rule.

All of this is significant insofar as it relates to the advent of what Simon Critchley and Jamieson Webster term 'a second late Elizabethan age,' which in essence means that within contemporary England policies of militarisation and surveillance are paramount to the conduct of ordinary life; one that is at once supported and contested within the scope of contemporary popular media (48). Within this highly mediated macrocosm it must be assumed that everyone is being watched and moreover, that 'everyone lives in fear because they *know* they are being watched' (50). At the same time, they exist in a state of distraction, as they try to live amidst a perpetual crisis of legitimacy installed to occupy their thinking as a complement to the paranoid reality of political life. Within such a schema our relation to the future is not one of being and nonbeing, but rather an eventuality that can be mapped on a continuum extending between absolute control and total submission. With some futures considered more viable than others, humanity's ongoing capacity rests on its powers of discernment amid a vast array of seemingly heterogeneous possibilities that seldom find their resort in the assignment of material value. Nonetheless as a closed system it does not excuse ignorance, nor abide loss.

In Elizabethan England, self-surveillance was but a mirror of a world that had become a prison replete with expansive and expensive networks of spies. The use of the Internet is simply a contemporary tool of the same historical set-up wherein the assembly of information, once more degrades into the assembly of commerce. Criminality and warfare have united to expand the use of military technology into an elaboration of domestic intelligence capable of dissolving the territorial ambiguities that once existed between nation and empire to achieve the universal generalization of warfare such that citizens and aliens alike become 'enemy combatants,' which must be wholly subsidiary to a security apparatus.

In order to understand the Elizabethan age, we must revert even further back to the time of feudalism, 'where the boundaries of civilization, dignity and hope no longer coincided with the boundaries of the nation' and state security has decentralized to include numerous governments, private firms and individuals (Hage, 18). The nation-state security provision thereby has reverted into uneven, and highly localized security marketplace organised by profiteering agents employed to establish a protective perimeter around urbanised life. 'As in medieval times, then, the result seems to be that the modern city emerges as what Holston and Appadurai (1999) have called a "honeycomb of jurisdictions", a "medieval body" of "overlapping, heterogeneous, non-uniform, and increasingly private memberships"' (13 qtd. in Graham 323). Within the confines of this second Elizabethan age, all forms of mobilization are designed to be somehow interrelated and permeated by 'the new tracking, targeting and access-control technologies of biometrics and pervasive discourses and tactics of securitization' (Graham 323). These assemblages are then aimed at commoner bodies deployed simultaneously as consumers, aggressors and informants.

During Elizabeth I's reign the Cecilian regime's relentless crusade for internal domination eventually produced a national consciousness, a mercantilist world economy, a concept of private property and a formalisation of Protestantism. The costs of their experiment in radical societal engineering was largely borne by a massive underclass, 'created by rapid population growth and combined with equally rapid economic changes,' who were ultimately put to use in the extensive wars dominating Elizabeth's last two decades (Breight 3). A nascent capitalism was to emerge from their sacrificial bodies, as the plunders of war significantly enriched the upper classes providing the capital for a system of oceanic trade and empire to expand into the seventeenth century. The violence these bodies endured were inextricably linked to the rise of a mercantilist global economy and the process of state- building in Elizabethan England that set an historical precedent for the aims of war and commerce to unite in a shared purpose; profiteering.

The plunder and predation that followed allowed private profit-orientated privateers to benefit at the expense of other actors — while being only superficially legitimated and directed by the Crown. The highly mobile and opportunistic nature of privateering allowed it to exist largely as a self-organizing structure that perpetuated itself through the violent seizure and forced appropriation of other vessels, their merchandise and on occasion their crews. Its situation emerged historically from a spike in the value of still a 'largely ungoverned space - the sea' (Elgoff 3). Through skilled manipulation of this unchartered medium 'quasi-state and non-state actors' effectively acquired the ability to exert 'significant influence on state interests and relations' (Elgoff 3). As merchant shipping increased, so too did the private capital to finance privateering ventures, which equally proved a highly lucrative enterprise. Mercantile companies that controlled privateering while primarily interested in unregulated profit-making, also performed a variety of critical semi-state functions including naval warfare. In England, these companies acquiesced to state power, while 'assuming sovereign-like functions abroad' (7). The state's capacities at this time were often insufficient or constrained to instigate the violence necessary to amass the wealth associated with powerful overseas expansion.

By contrast, the mercantile companies were able to craft and adhere to 'their own international policies, made deals with other companies or states, or were at war with them, engaging in open warfare, piracy, and privateering, sometimes independently and against the interests of their home states' (Elgoff 7-8). These companies ruled vast territories that far exceeded those held by sovereign entities. Through such authority they were able to fuse the interests and functions of politics and economics to establish trade monopolies. This is significant because technology companies and telecommunications giants, who have acquired vast market and informational power in and between countries, are exercising similar sorts of powers in a contemporary sense. Through establishing monopolies on the collection, access and distribution of information, these corporations as non-state actors have been able to expand their transnational reach at a time when the state is rapidly conceding its authority and responsibilities to private entities. The states enthusiasm for the cultivation and utilisation of private talent to secure its economic wealth and garner its defences amounts to a contemporary model of privateering, bound up with practices of both privatisation and sanctioned violence.

The Corporate Estate

Henry S. Turner in his book *The Corporate Commonwealth* speculates on the possibility of the 'a future "commons" being written *inside* the corporate form' as it now stands at the dawn of the twenty-first century (xiii). Turner makes the case that the joint-stock companies of Elizabethan England are 'the early modern ancestors of the commercial corporations that loom large over our own moment' (xvii). Then as now the corporation as a body, operates with distinctive regions devoted to 'gathering, recording, preserving, disseminating and archiving information as a form of capital' meaning that the corporation functions as a form of artificial intelligence, and 'as a technological *project* to borrow a term from commonwealth discourse of the later sixteenth century, that aims to assemble people and things' (xvii). While this body's primary concern is the materialisation of information, there remains a secondary concern for identity. In order to persist the corporation must generate itself as a 'group person whose head can emerge at any point, and who employs the hands, eyes, and mouths of its representatives which begin to move on its behalf and speak in its name' (xvii). In other words, the corporation must act as a body that is both immanent and functional in its capacity for recognition. Therefore, it is important to conceive that they ostensible purpose of the corporation; to process information, is only sustained insofar as it can remain vital to the transmission of commands across long distances. At their heart corporations must operate amidst an inherent appreciation of plurality and multiplicity. At their head corporations must operate through the utilization of 'not one but many heads, whose powers of perception can spread forth to the hands of its body, into the roving eyes of its factors and agents and in their mouths' (101). This in turn creates a "common" source of adopting values, as opposed to a purely elected one. The corporate body is one where, '"I" and "we" become interchangeable, singularity gives way to collectivity of the group person: the corporation has stolen the collective first person from the monarch' (101). If the monarch was God's flesh and blood representative on Earth, then the corporation is and remains to this day the facilitator of Man's elevation into a variable multitude.

It is helpful to consider that in its sixteenth century guise, what was thought of as the corporation encompassed 'The Church, universities, guilds, towns, cities, religious confraternities' as well as 'joint-stock companies,' (Turner "Corporations" 154). The corporation emerged to bring private capital to extend state power, at a time when the state found itself severely limited in its financial resources and looking to outsource its traditional obligations. Corporations were able to do this not only because of their superior access to capital, but also because of their increasingly superior access to information of both great economic and political worth. Eventually these corporations would evolve their own parallel structures of governance loosely keeping in line with those devised through state governance.

We are now entering into an age where large technology corporations such as Apple, Google, and Facebook are increasingly involved in acts of financial, territorial and ideological governance. Through establishing monopolies on the collection, access and distribution of information, these corporations as non-state actors have been able to expand their territorial reach. This is happening at a time when the state is rapidly conceding its authority and responsibilities to private entities. What we are now witnessing is the emergence of a new type of urban territorialisation wherein a plutocratic class of corporate celebrities bring their capital and class privilege to bear to establish a settler sovereignty on site of their headquarters, displacing and dispossessing the indigenous state sovereignty they have fiscally subdued to gain entrance.

These new corporate arrivals have little interest in articulating to the existing society, but rather concern themselves with creating their own, separate sovereignty within the local territories they encounter. This process is undertaken largely through grandiose settlement construction projects, which literally entail a clearing of structures, societal or natural, which came before them. The privatisation of publicly owned space is fundamental to this process, which is engineered to make sure that the local population find they have no right of place inside these meticulously fabricated commons, and their surrounding infrastructures are held in the domain of private ownership. Patrick Wolfe maintains that settler colonialism 'destroys to replace' and that expropriation in settler colonial contexts, 'is a structure, not an event' (388). The desire to establish private sovereignty over certain urban domains on the part of figures like Mark Zuckerberg, Sergey Brin, and Steve Jobs corresponds directly with the orchestrated bankruptcy of the state, as a direct consequence of extremely lenient corporate taxation policies at the local, state and national level, and lax taxation rules that allow for the illicit offshoring of billions in corporate earnings. By carefully annexing their wealth away from their locale of generation, these corporations have efficiently lowered their tax contributions across the globe to zero.

The rapid geographical spread of gentrification over the last decade, has been predicated on gaps in development on the local and national scale that have allowed speculative private developers to dominate and manipulate the progress of the housing sector. At the neighbourhood level, the poor and vulnerable often experience gentrification as a process of dispossession enacted by the more privileged classes, who have come to occupy the position of an imperial elite as they set up residence in desirable central metropolitan locations. The rapid appearance of this elite class in places like Menlo Park, Mountain View, and Cupertino is representative of global capital flow, which has created a state of exception wherein the new-found victims of employment restructuring instigated some years back find themselves now competing for the same territory as these more recent corporate settlers. Ironically, their employees will assume the position of either their dependents, or victims to come.

Through the expansion of full service corporate campuses, the generation of lifelong mechanisms of debt, and the simultaneous privatization of both the natural world and human data, it is possible to once again posit an emerging world of elite omniscience and mass servitude, where 'totalising power structures utilize contemporary information systems to enforce ideological homogeneity and quell resistance' (Kaufman 12). In exchange for security and protection those granted corporate tenancy, must agree to conspire with 'campus' rule, or risk career banishment. Through the intertwining of the trope of security with most everyday activities, the public is given over to accept 'risk aversion' as an enduring principle of reality and to bow those corporate employers seemingly best prepared to furnish it.

One of the final acts that Steve Jobs accomplished before his death was to secure planning permission from the Cupertino township council to build his master structure, 'Apple Park'. This awesome structure, a giant private compound shaped like a circle, would eventually hold 12,000 Apple employees. Jobs' amazing building was intentionally planned to appear as something alien, even potentially oppositional to the local landscape. Jobs observed, 'it's a little like a spaceship landed' (Jobs qtd. in Levy). Through this new headquarters, Steve Jobs was mapping the destiny of Apple itself as something that was designed to be futureproof, a structure that could not only survive climate crisis, but also equally stand as a stalwart of wellbeing - a Mothership exploring the limits of the great beyond of a corporate enclosure movement. Apple Park with its 'giant glass panels, custom-built door handles, and 100,000-square-foot fitness and wellness center complete with a two-story yoga studio', resembles nothing short of a manner born to reflect the specified vision of its master (Levy). What is not as apparent is that those there to serve the perpetuation of his vision, will do so at a cost of their own personal sovereignty and bond to a state that previously entitled them to certain freedoms of movement. Jobs' structure functions symbolically as a circle, where the past comes around again to take command of the future.

Jobs' ambition harkens back to an early modern concept of the corporation as a mode of organising collective purpose and systems of value. Jobs' neo-modern corporate political structure is mirrored in his last act of architectural dominion. His vision remains panoptical despite its break with a modern concern for a single point of view from which to maintain his cult of personality. Rather than use the luminous penetration of his mind upon the corporation he founded as a guideline, he had now commanded each of 12,000 employees to take up a concern for the company's continued interests. Through the promise of enhancing their own wellbeing they can express their subtle compliance with a multiplicity of observation, and beyond that an infinity of possible points of view. Apple headquarters are founded upon 1 Infinite Loop. With the advent of Apple Park, the matrices of vision are open wider than ever before, operating as a spatial prototype to mesh security with transparency. Its multiple points of entry, are flagged to correspond with various data points, involved in the mining of information within and beyond. Ultimately, this building reflects a desire to produce endless possibilities for employee convergence.

Jobs' compound's many internal elements mimic those of a prison or a barracks. Contained within it are recreation areas, sleep facilities and a mass capacity dining hall. And yet what we are seeing is not a model of disciplinary society based on Foucault's theorems, but rather something that appears to coincide with Deleuze's model of control - at least up to a point. Beyond that there is an inversion of the panopticon, to become something that has abandoned construction of the subject altogether, in favour of nurturing objective intelligence through *algorithmic* individuation. Within such a universe temporarily warped by a surfeit information, the deep state of the 'digital reintegrates the world into a *rendered* universe, viewable from all sides, modelled from all angles, predictable under all variable conditions' (Galloway "Laruelle: Against the Digital" 69). Here it becomes possible for a multiplicity of watchers, both human and machinic to be rewarded for their readiness to collaborate with Job's enduring vision, their multi-point aspirations made to converge now not within a single man, but rather a single entity. Such an architecture invites 'the possibility of enclosing the digital upon itself' and as such functions as 'an immunization against life' itself (Rouvroy). 'Within such an apparatus space itself curves to meet and comingle with seemingly limitless power directed against those who wittingly, or unwittingly transgress borders, harbour doubt, or express irreverence towards the given order' (Galloway 69). Apple Park's teleological universe demands that the subject cede his rights to alteration, to meaning, to 'human rights', or 'the right to separate free time from work timing and these kinds of things subjects could claim' that might disrupt the flow of radical standardization, the progress of *impersonation* (Rouvroy).

The universe Jobs' wished us to inhabit through Apple Park, is one rendered as predictable, but also perhaps more crucially, a universe that can be free of causality. One that is dependent on collective determination and a shared destiny through the conscious prevention of certain events from happening through pre-emption. Evgeny Mozorov refers to such a phenomenon as 'algorithmic regulation,' which adds up to 'an enactment of [a] political programme in technological form' ("The Rise of Data"). Giorgio Agamben discerns, 'if government aims for the effects and not the causes, it will be obliged to extend and multiply control. Causes demand to be known, while effects can only be checked and controlled' ("Theory of Destituent Power"). Within such a universe there is no need for progress, but rather for an ever more sophisticated presentation of distinction, division and the making discreet of the artifice involved in enacting power's proliferation. There is furthermore no tolerance for unchartered territory, other than its speculative territorialization.

Within structures like Apple Park, there stands a dialectical relationship between the territorial and the digital. Within them control over populations is maintained at once through what Helga Tawil-Souri refers to as ever hardening 'conventional borders' and the '"soft" realm of digital infrastructure' (Tawil-Souri qtd. in Dawes 6). Similar to 'the process of land enclosure, the privatization and monetization of information produces new forms of property rights and different systems of circulation, trespass, and exclusion' within the digital realm (5-6). Therefore, sovereign control must be exerted over both territory and technology in order to maintain a perception of spatial occupancy, as one reality no longer exists without the other.

States of Dissolution

We have entered into a new era in which the 'solutionist' cause-and-effect model has ceased to operate. The solutionist model was the archetypal model of intervention in the policy debates in the 1990s and early 2000s. During that time, it had been employed 'particularly around the legal and political concerns of the right of humanitarian intervention and regime change under the auspices of the War on Terror—operated on the basis of crisis or the exception' (Chandler 5). David Chandler asserts that its abandonment in the 2010s, speaks to a crisis within Western states to further act upon an understanding of their superior knowledge and thereby to effectively focus their political will. The state of the 1990s and early 2000s saw itself as an indissoluble entity that was 'timeless and could be exported or imposed—like the rule of law, democracy and markets' (6).

Subsequent to the financial crash of 2008, the state's stalwart self-image has been subject to a slow and painful diminution, overshadowed by the rise of what Evgeny Mozorov refers to as 'algorithmic governance' ("The Rise of Data"). Mozorov argues that it is the radical shift in authority, that hastened the death of politics as we have come to understand it. This new type of governance, brought into being alongside other neoliberal austerity measures, post-crash, forces political entities to strive to achieve an appearance of 'ultrastability' at a time where ultra-instability itself has become the default position of ordinary life. Governments, like their citizens, are repeatedly urged by corporate technology pioneers to focus their affective responses on the effects, rather than the causes of that instability. In so doing, responsibility is put on the state as well as the individual to become a better civic agent; one able to draw upon their limitless resourcefulness and resilience to overcome threats that seemingly come out of nowhere. Unemployment, debt, illness, and displacement that come to them as a direct consequence of the state's retreat its former position of governance are projected now simply as individual problems to courageously surmount.

The exception has become the rule when it comes to adapting to such crises, and preparing oneself to always do more with less. As the state copes with its own multitude of effects-driven crises, technology corporations stand waiting in the wings, with the aim to retool governance to resemble a 'lean startup,' whose operational watchword is optimisation. Their goal is to maximise citizen engagement and popularise themselves through their algorithmic analysis of their behaviour as consumers. Their priority of taming individuals, over taming the proliferating threats to human existence, should not be underestimated as we enter into this new era of cybernetic governance, where democracy is seen as a task of motivational management, over decisive influence. Such a condition obliges us to think about the end of democracy, and politics itself as a real possibility in our times. A cultivated desire for security has allowed those concerns to wither in favour of an absence of cares, and an apparent mitigation of risk as we allow ourselves to become enveloped societies of constant, supra-personal measurement and post-emptive valuation.

Apple Park's ascetic design sensibility borrows its conceptual basis from a world where intervention has come to be synonymous with the governance of effects. Such a design involves a shifting conceptualization of intervention itself. Giorgio Agamben has argued that 'since governing the causes is difficult and expensive, it is more safe and useful to try to govern the effects' ("Theory of Destituent Power"). The intervener, in this instance Jobs, is no longer is posited as the key agent for transformation. Rather, in an atmosphere no longer governed by the principle of cause and effect, sustained transformation is only made possible 'through the facilitation or empowerment of local agential capacities' (Chandler 3). Therefore, Jobs' role, and those of his tech contemporaries, is merely that of a leading light, a designer of motivation. Jobs' enduring influence over his company, can only be effectively projected if it is does so in a way that takes his employees away from 'the formal public, legal and political sphere' of the state and instead concentrates them into a seemingly 'more organic and generative sphere of everyday life' (3). Jobs' circular model of orchestration relies wholly on the careful management of effects to achieve the ongoing facilitative engagement in social processes within the seemingly endless span of Apple Park's interior logic. Build into that model is the constant subtly evasion of state government from the intended field of perpetual recognition.

This is what Antoinette Rouvroy is referring to when she speaks of the rise of an 'algorithmic governmentality,' as something that no longer concerns itself with imposing discipline on the conduct of a subject's life, but rather control and management of their every gesture of activity. As a consequence, 'algorithmic governmentality's' chief concern 'is not even a normalization of life anymore, it is a neutralization of life' (Rouvroy). Acting as a universal governing body, 'it immunizes capitalism and neoliberalism of all that could interrupt their course, of anything that could impose a crisis, or bifurcation' (Rouvroy). Agamben observes that 'starting with the Westphalia treaty, the great absolutist European states began to introduce in their political discourse the idea that the sovereign has to take care of his subject's security' ("Theory of Destituent Power"). In a subsequent era of algorithmic governmentality that concern has expired, and what has emerged to take its place in contemporary political discourse, is the idea that the sovereign has to take care of his data's security (Agamben). It is no longer other states, which threaten to impose a crisis or bifurcation of that ambition, but rather 'the world itself in its irreducibility to calculation, optimization and redemption' (Rouvroy). As a consequence of this, the notion of world in terms of geopolitics has been systematically replaced with a universe of big data. The world of Westphalia must be representationally exhausted and summarily obliterated from contemporary reality, not only in our conceptualisation of the past and the present, but most importantly the future, which must be totally 'replaced by a computable real made of digital flows' (Rouvroy). The dismissal of the world also necessitates a dismissal of subjects from the realm of political necessity. The universe of data requires this for its security, and thus is primarily concerned with the potential for machinic refinement, over human alteration.

Here it becomes helpful to envision the network that controls this transformation of the representational world into a rendered universe as something that is superficially imposing itself upon the subject's physical nature. In many respects, this mirrors the conditions of Christianity at a time when the Holy Roman Empire was ceding way to England's first empire, the Elizabethan Protestant Empire, which was in its own way was about coding and seizing control of interpretation through a contest of scripture. Sixteenth century England was considered a rogue state by Europe's great Catholic powers whose aim was a reformation of their previous Roman alliance of church and state. Elizabeth's appearance as head of both in her diminutive Kingdom was powered from an entirely different impetus. Hers was a zealous faith in information. Her legitimacy was not bolster by the divinely imbued blood that ran through her veins, so much as the cool organising intelligence that pulsed within her authoritative body and that of her government. Hers was to be a paper dominion, made real by maps and plans, documents and reports, great compilations dedicated to nothing so much as a proliferation of variously classified data that when assembled together gave an ascendency to her reign. All such matters were considered as an exchange of one reality for another, not as overt occupation, but rather as a veiled undertaking.

Dee's Lines

The study of geography at the university campuses of Oxford and Cambridge was a particular innovation of the sixteenth-century world. Its rise as a discipline coincided with the cultivation of a set of attitudes and assumptions that posited the English as separate from and superior to the rest of the world. Dr John Dee was at the forefront of this ethos widely held amongst Elizabethan geographers and cartographers. 'Dee, a Master of Arts from Cambridge, was a mathematician, astronomer, geographer, and on occasion necromancer' (Cormack 10). Dee became most well known as a learned and practical geographer and it was this aptitude above all that allowed him to direct the politically powerful men surrounding Elizabeth on a path towards imperialism. He is said to have been the first to use the term "British Empire" to describe the creation he critically guided into being. It was to be an empire of commerce, and not conquest' and as something built essentially to extend the English nation as distinct from the English state (Armitage 271). Leslie Cormack observes that 'the study of geography helped the English develop an imperial world-view based on three underlying assumptions: a belief that the world could be measured, named, and therefore controlled; a sense of the superiority of the English over peoples and nations and thus the right of the English nation to exploit other areas of the globe; and a self-definition that gave these English students a sense of themselves and their nation' (10).

In addition to his skills as a geographer, Dee was a highly skilled cryptographer. Dee's unique capabilities in this area, contributed to the development of an espionage network, which in time would become one of 'the most effective and formidable in Europe'. Established by William Cecil, under the direction of Francis Walsingham, this network came to rely heavily on codes' (Wolley 71). Dee's life was devoted to breaking open the codes of divine nature and he 'believed that numbers formed the basis of all phenomena and designed geometric algorithms to fathom the solar system. His students included Francis Bacon, promoter of the "scientific method", and the astronomer Thomas Diggs, who believed the universe to be infinite' (Jahme).

In the world that Dee inhabited there was no clear distinction made between science and magic. In the company of fellow Elizabethan spy Christopher Marlowe, Dee used the science of geomancy to conjure the starting location of the British Empire along Greenwich's ancient ley line, which he presented to the Queen in the fourth volume, *General and Rare Memorials Pertaining to the Perfect Art of Navigation*. Geomancy was a binary divination system that predated Leibniz's use of binary code. The site chosen by Dee to construct his New World was the Isle of Dogs, in East London. From its docks, a thousand ships would set sail to pillage, plunder, subjugate, and rule over numerous overseas dominions. Dee's mystical location remains to this day the centre of Britain's empire, this time financial, spilling out from the hermetic towers of Canary Wharf. Dee chose the site because it laid upon the remains of an Ancient British hillfort known as *Caer Ruis*, which he believed had once been the dwelling place of Merlin, King Arthur's chief magician. Dee proposed that a "British Empire," would be nothing less than a restoration of the reign of King Arthur, 'as he believed that Arthur's original colonies were, in fact, in the New World—even that America was Atlantis itself' (Louv). Dee saw Elizabeth as the living Arthur; himself, Merlin. Dee personally formed 'a company to colonize, convert and exploit the Americas by opening up a northwest passage that would eventually allow passage to East Asia, through the icy channels, bays and inlets north of what is now Canada' (Louv). The quest for the Northwest Passage was a key driver in the development of British imperial expansionism. The main channel of the Northwest Passage was legendarily anchored to Greenwich's powerful ley line that stretched all the way across the globe. 'There is strong evidence that Dee was the intellectual force behind Francis Drake's 1577-1580 circumnavigation of the globe' (Louv).

Dee believed that there was, '"an imperial formula" of mathematical simplicity and certainty' that would guarantee the success of Britain's new empire (Sherman 150). The formula was as follows: 'domestic and international security + territorial expansion = an "Incomparable *Ilandish Monarchy*, this "British EMYPRE" (150). Dee's imperial ambitions stretched centuries into the future. Equally they stretched far into the past to draw upon ancient and medieval precedent making the formula of empire balance at the fulcrum of retrospect and prospect. Imperial space according to Dee's geometric formulation is inherently non-Euclidean. It is dynamic, vectorial, transitional, or durational in its plural coordination of spaces. His formula allows for a world with multiple perspectives to come into view, where the virtual and the actual present themselves simultaneously, thus fragmenting and decentring the viewpoints that summon this new world into being. To his mind, it was only from acting within this dense, multilayered and overcoded space that one is able to reconcile universal truths with incontrovertible facts. Dee's desired empire existed both at the level of computation and transmission. By combining magnitude and direction into a mode of universal conveyance, Dee was convinced that he would be able to perceive the source of all occult wisdom.

Dee's ambition through his work as a polymath was to re-establish a communications network between the cosmos and the Earth through computation that would allow a new world order to come into being. Dee's expression of how this might work operationally relied as much on corporeal terminology, as it did networks or webs in which elements form interactive feedback loops. Such language would not appear unusual in contemporary computing discourse. In order to construct his formulas, Dee practised an alchemical geometry (what we now term n-dimensional geometry), which operated according to a logic of multidimensional space. Those pursuing this type of geometry were concerned with mechanical problem solving and sequential logic, but foremost amongst their concerns was the pursuit of an Ideal form drawn forth from a figurative fourth dimension. The culture of early modern occult scientists is not dissimilar to that of contemporary computer scientists insofar as their constituencies rely on a mutual understanding of 'a secret language', a reading of 'personal illumination books', and 'belonging to societies of initiates which are seen by the rest of society as wielding their esoteric knowledge' (LaGrandeur 174). Just as the Cabalistic combinations of the Hebrew alphabet and the various secret names of God literally flesh out the earth, 'the special codes comprised of algorithmic combinations of words, numbers, and symbols that today's computer specialists type into their machines actually weave together the fabric of virtual worlds' (176). What can be enunciated through 'early modern occult science could also be said of modern cybernetics, if one replaces "magic," "incantation," and "talisman" ... with "programming," "code," and "program"'(176).

In his enigmatic text of 1564, "The Hieroglyphic Monad", Dee constructed a visual symbol from common astronomical and alchemical symbols, which he believed could function as a new graphical means to understand reality. In this text, Dee presented the revelatory properties of his Monas symbol as so great they could now shed light on the laws of nature. According to Håkan Håkansson,

> The most striking feature of Dee's text is that none of the reappraisals he was laying claim to - of alchemy, arithmetic, astronomy, optics and of a host of other disciplines - were substantiated by referring to conditions in physical reality. These reappraisals were supposed to be achieved by contemplating the symbol itself. Rather than representing a knowledge which had been grounded in the world external to it, the symbol was the very means by which the world could be explored, ultimately revealing truths that no contemporary man had any knowledge of (78).

Dee's claims for the power of his Monas symbol derive from the Pythagorean teachings, which suggest a fundamental correspondence between man's employment of mathematics and God's creation of all things in the universe. 'According to the Pythagorean philosophy, the construing of the numeral system and the creation of the world could be envisaged as completely analogous processes, both having their basis in the same concept — the monad. As a strictly mathematical concept, the monad was defined as the originative principle of all numbers' (Håkansson 187). Dee's contention that the symbol was the very means by which the world could be explored establishes the monad as not only a critical agent to explore perception, but also as a fundamental unit of computation.

In the context of contemporary computer sciences, artificial intelligence, and autonomous systems the use of monadology persists as a crucial means to describe the emergent properties associated with formulating reality, i.e. the simple rules that come together to form complex behaviour. Perhaps behind these 'logic machines' are vast metaphysical speculations, as well as profound metaphysical truths that span the temporal divide between the Elizabethan age of discovery and our own. If we consider perceptible space as made up of a series of spiritual, rather than physical points there is no need of individuation between them. Therefore, knowledge doesn't need to be produced, so much as deciphered as something inherent to the monad awaiting calculation to make understanding surface almost magically, or almost spontaneously. One of Leibniz's favourite analogues for his monads is a point in geometry, at which an infinity of lines converge (Turner, *Early Modern Theatricality*, 353). These lines for Leibniz never settle on one point, rather they arise from an intricate crisscrossing of energies, 'looping vectors' which 'produce potentially infinite planes, glimmering in glimpsed parcels of space and time' (353). Gilles Deleuze observed that in Leibniz's thinking 'a world does not exist outside of the monads that express it' (Tge Fold 68).

Monads are distinguished from one another according to their powers of perception. The monads that are able to perceive changes within themselves most clearly are graded more highly than others who cannot adequately make such distinctions. The essential activity of monads is to mirror the objective relations of the universe from its own autonomous perspective. This unique point of view can be assembled with others to form an order of co-existence and together they form a population organised around the simultaneous perception of co-existing qualities and the order of successively perceived qualities. When these are multiplied an axis of time and space opens up such that each monad is immanently hinged to the point of view of the others and as they correlate with one another at the level of refraction of God's coding of them. It is that enlightenment that encodes information. Monads in themselves are non-physical and dimensionless, but when positioned as aggregate to one another act as the vectors that map and animate our world.

Forecasting Perception

Elizabethan England was an era of unprecedented speed when it came to both navigation and information. Hers was an age of data, wherein traveller's reports could circulate in tandem with the geographical inscription of new lands onto a virtually mapped reality of an England to come, through the means of nothing greater than information itself. Coupled with speed, communication bore a path to understanding that fascinated philosophers, military tacticians, and magicians alike whose formulas and equations acted as progenitors for thinking a universal wisdom, an 'online' language that could be imprinted on the world as logically as longitude and latitudes had. This language had to be artificial in nature, an encoded language that outpaced the need for translation and allowed an instant retrieval of higher forms of perception. The fate of the world would surely lie in the minds of the adepts who could be entrusted with a way to receive and classify such information. Such agents would have to obtain as it were access to supernatural, or non-organic powers of perception in the relay of information.

Telepathy was thought to be one means in Elizabeth's era, artificial language and cryptography the others, amongst the information-bearing arts. Cryptographers were the most pragmatic of this grouping in terms of their desire to achieve accuracy despite the necessity of interpretation logically attendant to their labours. Dee's intellectual interests ran the gamut from angelology and demonology, to numerology, cryptography and mathematics. Distinctions amongst them would have not applied in his enthusiasm for his interpretation of texts such as the infamous Book III of the German abbot Trithemius's *Steganographia*, which instructed its reader on how to channel information telepathically. Such a practice even at the time would have been considered diabolical in nature, and yet the text spoke to Dee as a means of achieving angelic communication. This suggests that cryptography as we understand it as a matter of informational statecraft in formulating strategy, diplomacy, and provocation, might operate at a deeper level to convey matters of esoteric importance.

What we are witnessing now, climatically with the desire to extract information and transmit it instantly and simultaneously as an object logged in the heavens, is the rise of 'cloud' computing. The Cloud both circumvents and circumscribes state authority. Benjamin H. Bratton predicts Cloud platforms will be the next system of state governance, through a situation where states move into the Cloud creating a kind of super-jurisdiction comprised of overlapping territories shared amongst a small number of elite corporations and states. Who they will govern reflects a move away from an ambition on the part of these entities to 'simply extrude national or ethnic identity online and re-establish borders there', or for economically, environmentally and politically displaced individuals to 'use online space to make irredentist claims to territorial sovereignty denied them on the physical ground' (Bratton 136). Rather, in Bratton's estimation, such claims of belonging would advance toward 'those that develop and form without [recourse to] geographic and historical legacy and for whom the *Cloud's* own geographic situation is the basis of emergent sovereign imaginaries' (136).

In this sense, the Cloud can be said to convey a new means of establishing citizenship as a fixed positionality 'structured around the slippery semantics of *User* identification and addressing' (136). This makes for a nomadic condition of sovereignty, which is localised but not delimited in its direction. No longer divisible by national boundaries, these individuals exist in 'a local absolute, an absolute that is manifested locally, and engendered in a series of local operations of varying orientations' (Deleuze and Guattari 382). The User in fact exists in two spaces that 'exist only in mixture: [a] smooth space [that] is constantly being translated, transversed into a striated space; [and a] striated space is constantly being reversed, returned to a smooth space' (474). This happens as the individual assumes a network and grows their relational intelligibility, as something akin to a monad.

Like the monad, 'the individual is important to the network only as a quantified self, which can be mined for data and information, and once that data has been harvested, it creates, through conditions of mobility and correlation, a data subject that becomes more important and has more currency and valence than the individual' (Shah 24). Nishant Shah describes this data subject as one that is 'atomic' and one that is 'produced from a series of actions,' who is graded according to its ability to give over information and extend its relationality as a generic entity (24). A new geography emerges based on computation to centralise the meaning of this subject, while at the same time burdening it with a new ontology mediated by algorithms, which separates the self from its former material and organic capacities, and joins them to a new protocol of infinite connectivity. This puts the data subject at the centre of the universe, but only insofar as one accedes to a universe that is fundamentally networked and digitised. This subject, stands only amidst a data driven society. This monadic subject must be by definition inward looking and must adhere to cartographies that conform with filtering, sorting, excluding and destroying as it travels the length and breadth of the network. Ironically, the data subject gets built up through a steady process of degeneration, decay, disorder and damage. It can only achieve mobility as a source of probable value when it begins to separate, to fall apart so that it can be then reconfigured into complex relation with all the other data sets.

Networked data subjects, 'like all digital objects', come 'into being because various databases connect with each other, and learning algorithms form simulation narratives between the different data sets, to create new conditions and forms of identity' (Shah 35). In this way, they are made synthetic and 'are denied the quality of autobiographical animals' (Rouvroy). Shah observes, that the networked data subject that emerges, 'is a mobile and simulated entity that has almost no relationships with the quotidian, affective, and subjective life and practices of the user' (Shah 35). All of those characteristics must be jettisoned in favour of a quantification of the self, made available to extraction as data. Only once that data has been harvested, do the conditions of mobility and correlation emerge to create a data subject. That subject then becomes more important and has more currency and valence than the individual' (35). Value is determined through the subject's accessibility, legibility, and intelligibility; all of which must be put into the service of systems information profiteering.

Cloud computing utilises a network of remote servers hosted on the internet to store, manage, and process data. Within that context, Etienne Turpin makes us aware of the concept of 'exocephalization.' This practice involves extracting 'the brain from the body and pulling it outside of itself into an infrastructure space'. The process allows the brain to be removed from the body, only to be uploaded into the cloud. This practice allows for the commercial appropriation of 'every aspect of social labour, cognition, invention, and innovation that capitalism would like to extract from each and every one' of us. This is where individualization and its persistence into a digital empire becomes realisable, insofar as each human being is credited with having its own special ideas, and innovative perspective on seeing things. What exocephalization achieves then, is a means through which giant technology corporations can 'move your thoughts into a place where they can be more effectively monetized' (Turpin). This cognitive displacement will be promoted as a victory of organization, when in fact it is a triumph of extraction, a novel form of primitive accumulation proper to the digital age.

While the Cloud may seem to imply a light and free as air method of information storage, its proliferation heralds the arrival of a new architecture of access, where ideas can be paid as rent to the giant technology corporations in a manner similar to feudalism. Within the structured domain of cloud computing, the lion's share of what users produce, directly and indirectly, goes on to nourish and edify the societal position of their lords and masters. Those in possession of large domains were in a position to profit from their user's vital dependency on continuous access to them. As such they occupied a position to determine the set of laws of how what was produced inside of these domains was ultimately valued. Here is helpful to recall the clouds of our second nature. By this I mean the second nature McKenzie Wark refers to as 'the space of the material transformation of nature by collective labor' ("Escape from the Dual Empire"). Wark describes second nature as a space of fragmentation, alienation, and class struggle, where nature is treated as though it was a standing reserve for industry to draw upon at will. As we enter into a third iteration of nature, it becomes impossible for us to obtain a vision of first nature, clouded as it were, by the dust of its own information as a virtual geography. The establishment of feudalism marked a transition from first nature, ultimately making way for a nascent second nature to appear in the form of early merchant capitalism.

Jussi Parikka reminds us that clouds with 'their seemingly light ephemeral nature are full of the chemical remnants of the on-going industrial age, what some call the Anthropocene' ("On Media Meteorology"). He cannily observes that 'many prefer to think of the current informational culture as one of light, marked by the weightlessness of fibre optics and the speed of digital transactions, and yet it is also one of weight – of minerals, metals, energy consumption, and entropy'. The Cloud carries this material as surplus to its production cycle. Whereas second nature was reliant on seascapes to provide transport for its endless flows of capital, labour, energy, resources, and products, third nature is able to realise things seemly out of thin air. The Cloud evaporates bodies, positing them into a new formula where they act as information hubs, attractors that conduct vast amounts of data. The Cloud itself maintains form through access to massive amounts of energy and to water for cooling necessary for the storage of interest within it, required to transport human worth in this rather more transient state.

Inhabiting Technology

We have now entered into a phase of existence where information now travels faster than people or things, this is what McKenzie Wark terms 'third nature', which summons into being 'a space within which all the possibilities for the organization of other spaces come together' (Wark 70). As this phase reveals itself, 'what this general populace takes to be a natural or naturalized set of circumstances, on which their day-to-day existence is dependent, is what Wark refers to as the 'peculiar geography' of a thoroughly mediated, vectorialized world (Coley 53). 'What the vector communicates is unknowable, a cloud of dust' (Wark, "Escape from the Dual Empire"). Trauma ensues in the instant the human world assumes recognition of itself *as* data and its basis for existence one fully subordinated to the software processes that 'tabulate, index and sort the entailed relations making up the human world' (McCarthy 133 qtd. in Coley 59). The acceleration of data threatens to relinquish the ephemeral modes of representation to ceaseless processes of computation. 'Phenomena that are massively distributed in time and space relative to the human, can be computed but not seen, experienced but not represented' (Morton 12 qtd. in Coley 54). This forces the human world to form relations with non-human agencies, to establish exchanges with the algorithms and other software processes that accelerate and intensify the possibility of human obsolescence in order to allow a profound upgrade of its ontological understanding to take precedent. Through a partial attunement to what is always already non-human in any human form of mediation to the great beyond, we are able to rediscover 'the actual inner logic of the age of intelligent machines', as at once 'the reason of trauma' and the instrument of catastrophe for a humanity still beholden to a linear process of rationality (Pasquinelli 8).

In the early modern era, 'practically minded "virtuosi" made their living by creating such devices for the princes and nobles of Europe, and Dee himself was said to have made a giant mechanical beetle for a play at Trinity College in Cambridge that astonished audiences. "And for these, and suchlike marvellous arts and feats naturally, mathematically and mechanically wrought and contrived, ought any student and modest Christian philosopher be counted and called a conjurer?" Dee asked' (Louv, "John Dee"). Dee's artificial creaturing technology draws its power from nature, his conjuring the 'augmented intelligence' of his day, similar to ours 'prevails by continuously adapting to its environment, encountering catastrophe and chaos with a sense of wonder and kinship' (Haraway 31).

Donna Haraway, in her famous essay, "A Cyborg Manifesto" recognised that 'human beings, like any other component or subsystem, must be localized in a system of architecture whose basic modes of operation are probabilistic, statistical. No objects, spaces, or bodies are sacred in themselves; any component can be interfaced with any other if the proper standard, the proper code, can be constructed for processing signals in a common language' (Haraway 32). Such a language is comprised of predictive possibilities too large for the human to compute and thus organic intelligence must be generated to mobilise the field of operations. The problem is that the human by itself is too large a data set to travel, and as a consolidated entity its attendant breakdown must be carefully arranged. The network then emerges as a means to constantly deconstruct the individual, directing its energy into multiple data streams 'which can be transferred and circulated across different routes, to keep the edges alive and the nodes activated' (Watts qtd. in Shah 35).

Antoinette Rouvroy observes,

> If what happens in the world disobeys to the pattern or if the prediction is proven false what happens it will never be understood as error, it will be understood as a further opportunity to refine the modelisation, a further opportunity for the machine to individualise itself...There is a huge competition now between machine individuation, algorithmic individuation, individuation of the machine, the learning of the machine, and individuation of people, which becomes just impossible.

Donna Haraway recognised as far back as 1985, that 'our best machines are made of sunshine; they are all light and clean because they are nothing but signals, electromagnetic waves, a section of a spectrum, and these machines are eminently portable, mobile—a matter of immense human pain in Detroit and Singapore' (Haraway 13). Humans to Haraway's way of thinking suffer because they are made of materials both solid and opaque, whereas cyborgs are far more ethereal in terms of their basic make up. Carbon versus silicon. Darkness versus light. Our better angels. God, who created all things in number, weight, and measure, arranged the elements in an admirable order, even if that order signalled our own demise as one of his creations so that he might achieve the completion of his order. 'Chaos was the term used by the ancients to denote the "confused matter and undigested mass" from which "the divers natures necessary both for the completion of the universe, and for the completion of the Philosophical Work [i.e. the alchemical transmutation of matter] are brought forth" (Håkansson 192). God is the bringer of the light, the agent of cryptic transfer, the universe his surface for life's transmission. Mirroring this endeavour is the silicon chip, which since the dawn of the information age, 'henceforth acts a surface for writing etched upon the universe a molecular scale; and whose progress can only be impeded by the catastrophe of nuclear annihilation. Even then, its demise will be one brought on by a cacophony of signals, 'by atomic noise, the ultimate interference for nuclear scores' (Haraway 31).

Returning to the refrain of lightness, is not chaos the reverse engineering of logic through the absence of enlightenment? If light can be used to encode information, then presumably, hearkening to the thesis of Dee mentioned above, vectors themselves can act as light, and in so doing take on the divine power to reproduce the world artificially as a mirror of reality. What separates one world from the other has to be the quality of sentience that organic life possesses that exceeds any attempt to impose regularity, uniformity and consistency on this plane of existence. Whereas the desire of machine learning is to replicate us on the level of function, ironically is logic itself that will undue its further development in that ambition. While it is true that the possibilities for subjectivisation are becoming 'incredibly rarefied,' according to Rouvroy, it remains practicable for people, 'not to do what they are capable to do, not to do what algorithms say they could be doing in the future... [for them to maintain] the capacity of reticence, of enunciation' - in other words of spontaneously *sounding off.*

In the new Elizabethan age of accelerated time we have gone beyond a concern for quantification to disaggregation. While the network may appear to us as something that is essentially a mapping device, it has little interest in tracking the circulation of humans, but rather information, and as such has no concern for the singularity of a subject such as Elizabeth I, or Dr Dee, or even God to enliven it. It derives its energy instead from a place of immanence, from signals, which are technological and acting on a basis of systemic perpetuation. Such a universe is in many senses folded upon itself and this sense it is, according to Leibniz one that is, 'fundamentally connected' (Marks 122). Someone (or something) with perfect knowledge-a God-would be able to unfold it all and see how each part connects to every other. In such a universe, there are not disconnected fragments but peaks of folds' (122). Information appears to always operate as a matter of 'folds within folds' and such was the case with the occult art of geomancy which was considered a supernatural art in the Renaissance, but nonetheless asserted 'influence on Leibniz's binary logic in the *Dissertation de arte combinatorial*. 'Leibniz was particularly interested in a series of hexagrams found in the I Ching or Yi Jing (c. 1150 bce) which expressed numbers in what appeared to be binary form' (Doroudi 2). The I Ching hexagrams, which in themselves were a form of geomancy, included solid and broken lines that progressed in a sequence that was unmistakably binary. For Leibniz, these diagrams, as a system of 'ones' and 'zeros' capable of expressing any possible value, reaffirmed his belief that God himself could create the universe from 'unity' and 'nothingness.' God essentially is conveying the whole of the universe to us and he is doing so in a code that is distinctly deterministic in nature.

Coding Exchange

Leibniz is usually credited with 'the invention of the binary language of Boolean algebra, on which the logic gates of computer circuits are based' (Marks 128). Leibniz was also a legendary Sinophile. Much of his understanding of the Chinese Empire was made possible through contact with Christian Jesuits who arrived there in the seventeenth century (Tao 60). Leibniz's concept of the monad as the essential form of the universe corresponds closely to the neo-Confucian concept of *Li*, which refers to the underlying reason and order of nature reflected in its organic forms. Leibniz was particularly fascinated with the Chinese language and its potential to act as a universal language, based on its use of pictograms and ideograms. Leibniz's contention was that Chinese was a superior language to English, because it was code and not sound based. In this sense it was cryptographic, free from the interference of both history and philosophy, it could serve a transnational interest. During Leibniz's time, Chinese was 'thought to have been invented all at once, in order to establish a means of communication between the large number of different nation states' that surrounded it (Tao 70).

Leibniz viewed imperial China as a dual empire with Europe, where the primary objective between them would be intelligence sharing. Leibniz lived in a transitional phase of European imperialism, when

Europe had already begun to spread its political and economic influence around the world, but had not yet established the pattern of formal imperialism that would make Europe a dominant world power in the late eighteenth and nineteenth centuries. While it is surely true that Leibniz interpreted China through the lens of his Jesuit missionary contacts and his own metaphysical system, he nonetheless saw their affinity to one another in terms of ethics, theology and metaphysics, as the potential basis for an Enlightenment super state. 'Therefore, Leibniz saw himself as best placed to reveal to the Chinese what was true in their beliefs and doctrines, while China and Europe thus brought together through this "commerce of light" could then civilise the rest of the world' (Kow 4). Centuries later, Leibniz's vision for a 'commerce of light' to emerge from a universal duality of language and mathematics continues to influence post-Enlightenment thought about how to optimise civilisation.

According to Wark, we are in a transitional phase now to another proposed dual empire, where 'the military-industrial complex of the cold war era has been replaced, not by a juridical empire of global law and trade, but by a new duality, a military-entertainment complex' ("Escape from the Dual Empire"). There are two aspects of this empire that compete in space for our attention, those that wish to transform information into a commodity and those that wish to transform it into a strategy. Like Deleuze and Guattari's concept of smooth and striated space, they overlap and contradict one another. Whatever form they assume, 'both are driven by the same imperative -- the vectoralization of the world' (Wark, "Escape from the Dual Empire"). The vector in this case, acts as a conjurer, 'it is what produces the world as such, as a space of property and strategy, a plane upon which things are identified, evaluated, commanded' (Wark, "Escape from the Dual Empire"). The language of the vector translates back into rupture, into chaos when it is stressed, denying its emanations the luxury of normality, of semblance. Rather they are affected by a pathology inherent to all forms of empire, but not identical to the empire that came before this current American articulation.

All sorts of components related to the psychic fabric of American democracy and society operate in the unique universe of forces, generating a distinct pattern for its supremacy is pitted against a communications breakdown within and amongst its larger entity. The data subject is therefore enlivened not through the protocol of any sort of Foucauldian biopolitics, but rather as a potent simulation of the citizen, through which an alternative horizon of politics can emerge. What is required of the data subject is that they give over their powers of interpretation fully to a network and focus their energies on invention. For it is invention that ultimately speeds the vector, generating potentially infinite spaces of emergence, co-mingling and transformation within the newest territories of cyberspace as well as intergalactic colonisation.

As data mining aims for an obliteration of truth and representation to serve the necessities of commodification and stratification, a material practice of communication is called for that expresses multiplicity without recourse to a logic of discreet identity and exchange value when it comes to the harvesting of human generated concepts, perceptions and affects. The space of invention has yet to be invented. What escapes the dual empire is the possibility of a material practice of communicating otherwise; communication as an expression of multiplicity, without the logic of identity that characterizes exchange, and without the logic of the other which characterizes strategy. If as Rouvoy says, 'the data are completely amnesiac of their production, of their conditions of material production, which are just hidden,' the human subject merged to its data set must be prepared to once again grant itself a history. It must prepare the way for a narrative understanding of its own production; one that denies sufficiency to power, while at the same time, inviting future possibility as an unknowable risk. The end of the future begins with an acknowledgement that every body eventually codes. It ends with the awareness that every body ultimately corrupts.

Chapter 2
Poverty and Experience: Walter Benjamin's War in the Age of Vertical Sovereignty

"We have become impoverished. We have given up one portion of the human heritage after another, and have left it at the pawnbroker's for a hundredth of its true value, in exchange for the small change of "the contemporary." The economic crisis is at the door, and behind it the shadow of an approaching war. Holding onto things has become the monopoly of a few powerful people, who God knows, are in no way more human than the many; for the most part, they are more barbaric...Everyone else has to adapt-beginning anew and with few resources." - Walter Benjamin, *Experience and Poverty,* 1933

"Many contemporary philosophers have pointed out that the present moment is distinguished by a prevailing condition of groundlessness. We cannot assume any stable ground on which to base metaphysical claims or foundational political myths. At best, we are faced with temporary, contingent, and partial attempts at grounding. But if there is no stable ground available for our social lives and philosophical aspirations, the consequence must be a permanent, or at least intermittent state of free fall for subjects and objects alike. But why don't we notice?" - Hito Steyerl, "In Free Fall: A Thought Experiment on Vertical Perspective," 2012

We live in social and political times that in many ways appear to radically depart from the era in which Walter Benjamin acted as a critical philosopher and cultural critic. But such an appearance of radical departure is to some extent misleading, as the particular issues we are wrestling with today are derived in part from the very systems and techniques of domination that Benjamin began to critique over a century ago. As we confront a critical terrain littered with concerns about neoliberal governance, neocolonial warfare, environmental denigration, the global preponderance of new forms of social media, the deepening of social and economic barbarity, and an increasingly violent political landscape, we must remind ourselves that what we are witnessing today is an intensification of the broad global trends that Benjamin sought to identify and understand.

Benjamin understood what fascism gave to the masses was the right to expression amidst a political atmosphere in which bourgeoisie rule remained intact. Therefore, they could be distracted by the superficial appearance of a new form of political life at the very moment that such a life was threatened with its abolition through warfare. Benjamin observed that 'war and only war, makes it possible to set a goal for mass movements on the grandest scale while preserving traditional property relations' ("Art in the Age Third Version" 269). The new technological resources of the early twentieth century provided a rich canvas through which to engage a new theatre of war resplendent in its novel apparatus of gas masks, megaphones, flame throwers and light tanks, which drew man into the centre of conflagration and in so doing redrafted humanity itself as a war machine. All around him would stand a new architecture that vertically elevated his gaze towards outsized 'armored tanks, geometric squadrons of aircraft, spirals of smoke rising from burning villages, and much more' (269). Within this scope man's productive forces could be accelerated to produce destruction, and technology made the supreme tool of self-alienation. Within such an aesthetic field it is possible to ignore the real causes of unemployment and lack of markets and pleasurably assimilate the progressive annihilation of humanity itself.

Such gratification comes to humanity by way of a new technology; the newsreel, that summons viewers to peer into a bed of trenches, awe of incendiary bombing over cities, and grimly anticipate the effects of gas warfare as an entertainment of the highest order to be projected onto the consciousness to habituated audiences globally. It replaces the thoughts one might want to think with moving images through a process which radically disrupts the train of association, shatters the track of contemplation and startlingly redirects one's attention towards an onslaught of new information. The newsreel then can be said to act as the 'true training ground' for new forms of human assembly to come to the fore best suited to the camera (Benjamin "Art in the Age Third Version" 269). The pervasive interpenetration of reality by the filmic apparatus granted the masses the ability to reproduce and record the violent and dynamic quality of contemporary life. In so doing, they modified their own comportment to mimic the motion and force of the industrial machine and modern warfare. The newsreel allowed for 'great ceremonial processions, giant rallies, and mass sporting events' and global warfare to be 'fed into the camera', in order to simultaneously empower 'the masses to come face to face with themselves', only to misapprehend their growing standardisation with a revolutionary movement (282 footnote 47). Fascism's aerial perspective made it possible for war to be construed as accessible and permissible and for truth to be made profoundly distant from itself.

From that skewed vantage point 'mankind prepares to survive civilization,' allowing itself to be appropriated at will and claimed as the property of mechanization (Benjamin "Mickey Mouse" 545). Benjamin was convinced that the filmstrip shared a natural affinity with the assembly line. Both came into being at roughly the same time and had come to play a pivotal role in allowing capitalism to disaggregate and dissect the expressive movements of humanity and reorder them to conform to the laws of motorial functionality (Benjamin "The Dialectical Structure of Film" vol. 3, 94). The route taken by humanity in such a circumstance 'is more like that of a file in an office than it's like that of a marathon runner' ("Mickey Mouse" 545). Through such a route a false image of the past becomes indexical. The filmstrip's principle realism reassures its audiences that 'the truth will not run away from' them, as its movement can now be tracked and recorded (Benjamin "On the Concept of History" 390-391). Yet it is in that very instant that the true image of the past flits by, plunging them further into darkness. Roughly pressed against smooth glass, the filmstrip's purpose is not to fix the image, but rather to penetrate it and force it to relinquish the confidences and belongings congregating within it. The filmstrip's non-binding quality speaks to a new poverty of experience in which impression is erased, and in its place apprehension is hardened. In the process, the subtle traces of a life are made typical, aligned within a banal sequence and bureaucratically framed.

Audience Testing

Benjamin observed that through film, an ordinary individual is made to understand, 'that his performance is by no means a unified whole, but one assembled from many individual performances' (Benjamin "Art in the Age Second Version" 112). His performance ultimately will be judged by the masses whose invisibility heightens their authority of control over him. In seemingly apprehending the individual, rather than the class to which he belongs, it is possible to convince the masses that 'any individual can be in a position to be filmed' (114). This sets him up from the start to be examined and intervened upon if necessary without his consent; moreover 'these tests are performed unawares, and those who fail are excluded' not only from the stage of social relations, but from their own persons (111). As a failed social actor, 'he can be stripped of his reality, his life, his voice, the noises he makes when he moves about and can be turned instead into a mute image that flickers for a moment on the screen, then vanishes into silence' (112). His final resolution eerily presages a comprehensive liquidation of certain elements of the world's population, by transforming them from subjects into objects. These elements become discarded objects and useless commodities dimly occupying the conscience of a traumatized audience already intimidated at the prospect of their own material incertitude. Hence for Benjamin, 'film is the art form corresponding to the pronounced threat to life in which people live today' (132 footnote 33).

While the filmic apparatus can dispel the social actor, what allows him to be operational in the first instance is the aestheticisation of the political. Benjamin observed that 'Hitler did not accept the title of president of the Reich; his aim was to impress upon the people the singularity of his appearance. This singularity works in favor of his magically transposed prestige' (Benjamin "Hitler's Diminished Masculinity" 793). Benjamin's statement reveals that Hitler is not a politician in any traditional sense, but rather a new type of figure generated from a sensory infrastructure engineered to transform an older, formal act of hearing into a mass expression of attention governed by reaction. Through the newsreel Hitler's national audience are summoned to bear passive witness to the 'deformations, stereotypes, transformations and catastrophes,' which materially assail them in the world in which they live in. Their advent is masterfully sequenced to correspond with various 'mass psychoses, hallucinations and dreams' that thrive amidst the strategic bombardment of their 'collective perception' ("Art in the Age Second Version" 118). Hitler's command and control of the masses relies on the collusion of information and entertainment that had developed during the previous century. He then adds to it the most recent technological advances for managing complex systems to extend tensions beyond of the boundaries of the military and industry, to order and captivate civilian life. Hitler's power lies in his ability to 'create figures of a collective dream, such as the global-encircling Mickey Mouse' able to deliver on his sadistic fantasies of global domination (118). In such a scenario, personal autonomy must be constrained and individual ambition fulfilled from without, with an eye toward the greater essence of a society that obeys commands. If one takes initiative within the scope of the collective he does so as an expression of violence.

Under the terms of Nazism, progress itself is prefigured as a storm that violently disrupts the normal assembly of politics and installs in its place the spectres of fear, attraction and discipline. The force of that storm was captured endlessly through newsreels making the image of the relentless forward march of jackbooted troopers synonymous with Nazism itself. The filmic apparatus positions the subject as a point of view, as spectators willing to casually accept mediation as one in a number of battlegrounds they would be forced to confront in an ideological terrain seeded from the possibility of the extinction of those units of human capital that failed to credibly conform to standard. The filmic apparatus would bring spectators closer to the exercise of state violence than at any other time in history and to apperception of the hapless creatures for which the state remains a repressive apparatus of violence above all else.

The violence 'directed against political or social "trouble-makers"', according to Detlev J. K. Peukert, 'was not only not concealed from the population -as many who pleaded for excuses were to suggest after 1945 - but was highly visible, was documented in the press during the Third Reich, was given legitimacy in the speeches of the Reich's leaders and was approved and welcomed by many Germans ...while its targets were 'enemies' on the left and later the 'asocial' (197). German society's indifference to its growing violence was derived from the 'systematic destruction of public contexts and responsibilities and their subsequent dislocation of social forms of life' under the Third Reich, which made critical inroads into 'traditional environments, which provided some measure of refuge and scope for resistance' (241). Relationships associated with the private sphere during this era would be drained of their vitality and that energy redirected into the public demonstration of support for the Reich. Nazi Germany would hence emerge as a virtual community, one that coerced isolated and self-sacrificing modes of participation, and repelled all potentially dangerous social connections and meanings that might undermine it.

Germans were to consider themselves as actors playing a part in an historical epic, where their individual likeness would be captured struggling heroically in battle and later projected as a big budget film to be appreciably consumed by future audiences. 'Only a few weeks before the German capitulation, Joseph Goebbels used the premiere of the feature film *Kolberg* in April 1945 as an opportunity to hammer home the credo of his unique approach to politics once more. "Gentlemen, in one hundred years' time they will be showing a fine color film of the terrible days we are living through. Wouldn't you like to play a part in that film? Hold out now, so that 100 years hence the audience will not hoot and whistle when you appear on screen"' (Koepnick 51). Goebbels' remarks are based on a total aestheticisation of the contemporary situation in Germany in 1945. He addresses his fellow countrymen as though they were the 'living dead,' posthumously recorded as actors in the narrative of history being manufactured within their lifetimes. Having become accustomed to inhabiting a world of artificially induced images and sensations programmed into them by figures like Goebbels, they have been willing up to this point to reject the diversity and intricacy of global relations, in exchange for a monolithic system that persecutes independent thought. Nazism, then in some senses operates, as its own projector able to thrill and terrorise its audiences into an acceptance of a totalitarian system. Its main actor, Hitler, at its helm performing the role of an icon of divine omnipotence and absolute moral legitimacy made proper to both modern technology and industrial capitalism.

Deploying Media

Lutz Koepnick, in his reconsideration of fascist aesthetics cannily discerns that 'Benjamin--confined to the condition of exile--primarily deciphered the politics of fascist culture from "above."' (54) In this way, he is forced to act as his own angel of history, an apparitional figure hovering over the various types of cultural fragments at his disposal 'to decipher how Hitler's subjects inhabited both the political spectacle and the symbolic materials of a modern leisure and media society in order to take position and construct their identities, however precarious and inconsistent' (54). Benjamin is able to conduct his research from this position, because Koepnick observes, 'fascism constitutes a phase of capitalist modernization in which the political dimension itself becomes a market item' and therefore is 'circulated as one of many other objects of popular desire' (55). Nazism therefore can be said to operate as a brand name replete with its own trademark, the swastika, which allowed it to operate universally in much the same way as Disney's mouse. Similarly, its success depends on cultivating mass loyalty and consumer satisfaction. Nazism's plan for market domination hinges on its ability to fabricate dreams and disseminate delusions that mimic the appearances of social reality and at the same time undermine it. National Socialism, therefore, can be understood as a monopolistic enterprise, insofar as its brand of 'fascism constituted a dictatorship *over* the new media and a dictatorship *of* the new media' (Koepnick 62). This vast range of influence allows it simultaneously to promote images of national unity and communality, and in reality, produce solitary, isolated individuals that were made ready to consume warfare, as one in a number of ordinary rituals associated with daily life, leisure and consumption.

Benjamin's montage reading of history takes place amidst a crisis in democracy, wherein mass media has triumphed by transforming politics into a screen and history into a theatre of war mediated by technology.

Filmic imagery functions as a temporal index that enables fragments and moments of the past to intrude upon present experience. 'Such images "preserve" experiences', which can be re-enlivened and redeployed elsewhere in history in aid of 'the possibility of colliding with their present and future viewers' (Lindross 238). In the twenty-first century that screen has split off from itself and proliferated onto multiple surfaces. As a result, contemporary political power lies in the control not of the means of production but of information. Thus, it is possible for an inverted image of totalitarianism to flash up at this point in history, and be altered to support our present condition of perception in distraction.

Goebbels' premonition came to pass in countless filmic renditions of those terrible days. Ironically, it is those that did not resist playing their part that are harshly judged in cinematic perpetuity. The depiction of the Third Reich now makes up the stuff of perverse enjoyment for postwar audiences, who never fail to delight in witnessing the villainy of Nazi characters, their exegetic defeat figured as a pleasure of the highest order. Perhaps the greatest lasting achievement of the Third Reich is not a question of their celebrity versus their infamy, but rather their radically reshaping of common standards of decency 'to render aesthetic pleasure a direct extension of political terror' and introduce 'a form of violence in the service of future warfare' bearing vast implications for our contemporary fascination with total destruction (51). Goebbels' provocative association of consent with self-discipline through filmic participation demonstrates film's escalating role in the governing of mentalities as the twentieth century progressed.

Time and space were radically reimagined to play out in Benjamin's work in cosmic dimension and introduce relativity into the equation of perception to correspond with a sophisticated acknowledgement of warfare, advertisement, and advanced industrialisation. Benjamin's appreciation of these developments as crucial to the formation of a modernity, in many ways anticipates our contemporary visual field as one saturated by militaristic and celebrity imagery integrated into a globally mediated milieu. The intensification of this process in the twenty-first century creates the atmosphere for a new subjectivity to emerge which allows itself to be safely enfolded into an order of being that coerces its acquiescence to surveillance technology and screen-based distraction. One could argue, that such an order contains within it, already the germ of radicalization insofar as it forces the subject to be 'exacerbated,' and by definition, intruded upon 'by the one-way gaze of superiors onto inferiors, a looking down from high to low implied by these new formations that massively extend the societal pressure imposed on individuals' (Steyerl "In Free Fall" 24). Added to this is the tremendous weight of information solicited from these seemingly banal, technologically mediated encounters, which assiduously erode the boundaries of warfare and entertainment.

A post World War II world witnessed an amplified version of the portrayal of political and social violence, both in society and its representation on film and later television. In the twenty-first century, digital technologies not only delineate the screen actor's actions, directly or indirectly; they also function as a microcosm of the societal field of action through which the audience is compelled to act. In turn, the characters modelled on these actors are carefully depicted to demonstrate the control exerted by the political apparatus itself. Audiences superficially appear to care about the depictions of various peoples, or politics, however on a deeper level they are concerned about how they themselves are performing in any public guise, or acting on behalf on any commercial, political, or social faction. In both instances, they are being instructed that they have to be good enough to dominate, or at the very least, avoid being dominated and humiliated by other people. The panoptic effect of the digital recording of all actions in public space grants the actor either greater social capital as a reward for his conformity, or greater social censure as punishment for his variance. Regardless, each individual is made tacitly aware that his performance is under constant, subtle scrutiny.

Film as a technology was in its absolute sense always meant to be a tool of apprehension. Its assimilation into mass culture was made possible, because the majority of city dwellers working in offices and factories had been previously made to relinquish their humanity in the face of an apparatus' prior to its invention (Benjamin "Art in the Age Second Version" 111). Its entertainment value is a secondary affect. According to Benjamin, 'collective laughter, is one such preemptive and healing outbreak of mass psychosis. The countless grotesque events consumed in films are a graphic indication of the dangers threatening mankind from the repressions implicit in civilization' (118) The filmic apparatus is the 'plowshare that cuts through the masses' loosening them up through laughter, only for them to be then 'stamped down hard and firm' into the soil of an emergent Third Reich, which provides for no possible uprising (Benjamin, "Hitler's Diminished Masculinity" 792). Benjamin perceived Hitler's image as 'not one of the military man, but rather the man in easy circumstances', refashioning the feudal emblems of power to appeal to modern audiences, whilst at the same time reconstituting the rule of the bourgeoisie (793).

For Benjamin, the "mass," like the film, is "torn apart" (zurstückelt), scattered, strewn in pieces that must then "be assembled in accordance with new laws"' (Rutsky 14). Film itself emerges as a sort of terrorism insofar as it "explodes" the everyday world, stressing reality to the limit in order to expose its contingent, dispersed, and fragmented components that anticipate their digital reenactment. Lodged within this new principle of reality subjects and objected become inseparable from one another as filtering devices for various modes of appropriation bitted out so as to be on one hand, 'redirected, re-embedded within the fabric of traditional commercial and property relations, which are generally couched in the rhetoric of consumer choice and self-expression' (19). Informational flows themselves are tethered toward their commodification as a time when labour is increasingly made free of value. Equally it's assumed that information as a property, information must be given a proper address and a destination, while its receivers are forced cohabitate amongst a surreal landscape fraught with discontinuity and dispersion and baited with irrational cuts, aberrant movements, and false continuities.

Vertical Domination

Hito Steyerl asserts that 'our sense of spatial and temporal orientation has changed dramatically in recent years, prompted by new technologies of surveillance, tracking, and targeting' ("In Free Fall" 13). There are undoubtedly consequences to this proxy form of linear perspective 'that projects delusions of stability, safety, and extreme mastery' as though they are causally related back to the sovereignty of nation-states (13). For all their efforts, they also need to create societies convinced of their perpetual decline as evidenced by their sinking further and further into the urban abysses and splintered terrains of multicultural occupation, surveilled subversion and policed biopoliticality. Within such a scenario society itself has reached 'terminal mass,' its passive condition relayed to it through an admixture of uncontrollable consumption and panicked hyperconformity, its treatment 'atomized and dispersed [to it] at the end of communication receivers' (Terranova 136) Through a potent combination of income insecurity, social breakdown and environmental denigration, society has been starved 'of its revolutionary power in a kind of entropic dispersion' bent on laying waste to the overall system of meaning (136).

Digital technology has made it possible for contemporary audiences to transport themselves at will over virtual landscapes of military domination and stoke their fantasies of power through worship of hyper-masculine, atomised, technologically hybridised warriors fighting to defend 'our way of life'. It has liberated them to aspire to eternal youthfulness through plastic surgery, gruelling fitness regimens and other celebrity inspired self-improvement techniques to maximise their individual value. It has encouraged them to find satisfaction in recording their banal physical actions measured per second in a bid to make it to stardom on YouTube, or Instagram, and therefore do their bit to maintain a dream-laden culture of ever-expanding digital capture and mediated recognition. Digital technologies have become agents to bring reality itself into being. Equally, these technologies have assumed responsibility for conveying history, as well as beautifying it. Operating as reality's plastic surgeons they make it possible for any blemish to be gradually wiped away. So too the scars of history are obliged to fade away from our collective conscience as the gross imperfections of a thankfully, bygone analogue age.

Steven Shaviro postulates that we now inhabit a post-cinematic age that intensifies the shock effect Benjamin described several decades earlier. Today we are dealing with an inconceivable complexity when it comes to new media formations, which constantly pervades upon reality forcing the global population to imagine their acceleration as the only future possible. 'Just as film habituated the "masses" of Benjamin's time to the shocks of heavy industry, and dense large-scale urbanization, so post-cinematic media may well habituate Hardt and Negri's "multitude" to the intensities arising from the precarization of work and living conditions, and the unleashing of immense, free-floating and impersonal financial flows' (Shaviro 138). Neoliberalism requires an emptying out of capitalism such that imagination itself has become the target of an ideological austerity and its systemic failure the only coercive option left open to the "multitude."

Benjamin's aim was to blast through capitalism's systemic production of the image and sound, so that the material contradictions and possibilities of its articulation can be made real other than as a labourer carrier of information. Benjamin calls for a break in the action of producing meaning, an interruption to the act of acceleration such that catastrophe and destruction can be bypassed, tracked elsewhere, at least temporarily, so that humanity might get the opportunity still to 'rethink production and the very notion of "productive forces"' (Noys 92). Benjamin's desire to slow down history is directly at odds with his Marxist counterparts in the Frankfurt School, who are convinced that the only way to harness the energy of revolution is to tap into the powerlines to capitalist production. This ideal has inadvertently given capitalism a monopoly over the energies associated change and the future, the consequence of which is a politics of deferment, wherein we as a people must always exist in a 'state of emergency,' as opposed to a state of realisation.

In Benjamin's essay on Surrealism, he refers to Breton's 'intention of breaking with a praxis that presents the public with the literary precipitate of a certain form of existence while withholding that existence itself' (Benjamin, "Surrealism" 217).   To grapple with such withholding is the essence of Surrealism's poetics inventively comprised of 'demonstrations, watchwords, documents, bluffs, forgeries...' (208). Their poetry is born of a residue through which it is possible to see perhaps for the first time, how political and factual reality is technically organised, and therefore can 'only be produced in that image sphere to which profane shock illumination initiates us' and presents us with its commands (217).

What the Surrealists were able to achieve at least temporarily, apart from their leftist counterparts, was to at once realise the exchange value of man's features for those of the machine and to make that the cause for an alarm to continuously sound: 'They exchange, to a man, the play of human features for the face of an alarm clock that in each minute rings for sixty seconds' (218). The sensory jarring produced by such artificial mergers permitted Benjamin to 'imagine an alternative reception of technology' that could introduce a mode of piercing reception that might penetrate 'the scar tissue formed to protect the human senses in the adaptation to the regime of capitalist technology' and therein recover humanity's innate sensory affect (Hansen 308).

We might refer to this malady as a world-system end-state where the masses are locked into a common informational milieu and the greatest penalty to be exacted, is excommunication from the self-same realm of capture. Faced with this threat, the terminal mass accepts its daily assault of images without context and indignation without address transmitted relentlessly across a hyperconnected planet. Tiziana Terranova explains, what distinguishes the terminal mass from its proletarian predecessors is that, at its receiving end 'it cannot be made to form a stable majority around some kind of average quality or consensus,' (153). Rather, it has become fundamentally acentred, 'under the recombinant assault of informational flows' that have succeeded in compromising the wider field of spectacle, so that it can no longer be 'made to unite under any single signifier' but instead amidst a virtual space that is 'common, without being homogeneous or even equal' (154). Hence, the internet (what we used to refer to as the world wide web) 'gives visibility to a larger feature of our communication milieu – and one that is on its way to becoming hegemonic, as all communication systems become ever more interconnected' (154).

There is nothing inherently redemptive about this new integrated configuration, which demands of its participants a willingness to engage with a self-perpetuating showground that decentralizes their interests, distributes them amongst differential communication capabilities, and commands them to develop an emotive political intelligence honed within the terrain of the common. Such a phenomenon, heralds nothing less than the emergence of a new form of linear barbarism set on 'making time as smooth and digital as space', therein lending privilege to the vertical axis as the means through which to observe humanity's spiral downward, where the contours of imagination are at once suspended and enthralled by 'the elevated view, whether of the satellite or the drone' (Wark).

The contemporary drone operator functions in a strikingly similar manner to Benjamin's "Destructive Character" in that he has 'few needs, and the least of them is to know what will replace what has been destroyed,' nor must he concern himself with 'the place where the thing stood and the victim lived. Someone is sure to be found who needs this space without occupying it' (Benjamin "The Destructive Character" 541). As a technician, his function is 'to pass on situations, by making them practicable and thus liquidating them' (541). What has previously existed must be 'reduce to rubble - not for the sake of rubble, but for the way leading through it' (542). Amidst the free-fall of this early twenty-first century scene it is possible to conjecture that the disordered aftermath remaining before him of 'living and non-living things might be a heap rather than a hierarchy' (Steyerl qtd. in Wark). The rubble mounting upon itself might eventually give way for a new order to emerge, based on a scarcity of leverage. Here the only advantage left 'is to be on top of the one falling below, a kind of 'vertical sovereignty'" (Steyerl qtd. in Wark).

If the previous century managed to transform 'the earth from a continuum of places to a grid-like space,' as McKenzie Wark argues, this one is determined to dissolve such fix coordinates altogether, in order to force mobility and level velocity. Nevertheless, Peter Mörtenböck maintains that the whereabouts of these orders and spaces can be 'precisely calculated' by the 'apparently boundless mobility of labour power, worldwide data traffic, the uninhibited flow of capital and the proliferation of hybrid lifestyles' (107). All of these formations correspond with an intensified class war from above provisionally discernible through 'the flight of business and politically motivated expulsions,' which compel 'new worlds to emerge whenever these flows are bundled - at the numerous nodes of advanced capitalism' (107). Subject to intense friction, the landscapes of lateralized market power coalesce around 'a set of cultural practices and interactions between particular emergent assemblages' that have as their basis the postcolonial world order, and the production of transnational spaces (107). What distinguishes them from one another is precisely their 'vertical sovereignty', evident at every level within neoliberal societies from 'the towering heights of investment banks, hedge funds, and state-run sovereign wealth funds to the lowest depths of … the global "creditariat"' (Haiven 94).

Within this financialised hierarchy it is possible to observe the articulation of a range of trans-territorial spatial formations, which are able in turn 'to fuse multiple politico-economic interests with the process of subject formation' (Mörtenböck 107). This fusion generates a radical altering of what we understand as citizenship under the traditional terms of sovereignty, and reshapes it to conform with 'international corporate interests, the differentiated exercise of state power and the contingent struggle of citizens themselves, thereby extending the concept of citizenship beyond the idea of territorialized rights,' toward assembled rights made up of any number of 'strangely unresolved hybridities' (108). Within such a schema the individual's profound lack of territorial consistency is reflected in the way citizenship is offered to him as an individually assembled package whose various, formerly legitimating components are not in any definitive way guaranteed to him. The seemingly permanent erosion of stability experienced today as an aspect of citizenship corresponds with a pattern of unceasing global migration 'aligned to the quid pro quo principles of market relations' which have now been recalibrated to act as modes of subjective authentication (108). Mobility is the only constant within such a framework for self-understanding, and relies heavily on one's willingness to compulsively engage with economic discourse that makes self-worth synonymous with accumulation and the assimilation of market values.

Under the new terms of such citizenship the individual is compelled to enter into relations with the state, not at the level of territory, but of wealth, prompting citizenship to be entered into as a contractual agreement whereupon the citizen accedes to inclusion based on criteria that conform to standards of 'economic viability, efficiency requirements, competitive pressures and terms of trade' (Mörtenböck 108). Each applicant is judged from the perspective of projected earnings and their likelihood to sustain economic viability within a context of relentless economic crisis. The would-be citizen must be able to constantly absorb risk, whilst at the same time adapting to contingency. Understood in terms of verticality, Mörtenböck argues that migrant subjects now exist in a state of multiple graduated legal situations and as such are forced to occupy 'grey spaces' where their status is kept at the level of interim, and their appearance within the nation-state one of perpetual ambiguity, wherein their acceptance or rejection is forced to appeal to an unstable political geography. The obscurity of their condition orientates with the blurred boundaries of a former colonialism, and at the same time points the way towards social and spatial relationships in the present that are informed by the emergence of powerful state and private actors tasked with creating new zones of exception through which to administer judgment of desirability, toleration and criminalization to these weak and marginalised communities. Within such a formation it is possible to directly connect the market orientation of social relations to the generation of interminable conflict zones.

The 'grey zone' that the 'so-called 'Islamic State' draws upon in its recruitment strategy, may be construed to reference to this new constituent terrain of Muslims living in the West who dwell in state of social alienation and cultural disaffection. In the Western world citizenship increasingly breaks down along lines of racial profiling. Those visually perceived as of North African or Middle Eastern descent are classed simply as 'Arab,' or as 'Muslim,' and thus rarely distinguished as fellow members of a national community. These same individuals find themselves estranged from what are assumed to be their own cultures and mores, and from mainstream Islam, as they are from wider Western society. With no reliable milieu in which to situate their belonging, they find themselves within a space of cultural voidance that demands its own form of expression, which harbours within itself the potential to later configure as an attraction to Islamism and fascination with jihadi-style violence. Without recourse to any spatialised cultural coordinate, they become driven to disrupt society from a tele-temporal perspective, offering their own alternative forms of cultural dissembly. Terror becomes a perfect device for destabilising their observer's perspective and breaking down linear time to re-orientate it to a large extent. It demolishes the Western perspective causing it to disintegrate into different types of abstraction, hence the so-called war on terror.

Mediation of Civilisations

In the Middle East and North Africa, the rise of Islamism runs parallel to the United States' fundamentalist assimilation of neoliberalism. Both phenomena rely on a depoliticisation of the economy, and a dismantling of the principle of democracy, in aid of evacuating the grounds from which to confront real social conflict. This effectively annuls the basis from which any legitimate form of socialism might arise, replacing it with a culturalist logic that mitigates 'the perversions of actual political and economic grievances,' effectively transforming 'struggles from real social contradiction to the world of the so-called cultural imagination, which is transhistorical and absolute' (Amin). In order to facilitate this shift, 'whole economic histories needed to be rewritten' to allow 'the pre-eminence of the United States and Great Britain' to be 'posited as a triumph of laissez-faire economics over lazy French interventionism, rather than a bloody hundreds-year-long process of redesigning the world by force of arms at the expense of entire civilizations' (Khalil).

The ascension of neoliberalism depended not only upon eliminating alternative understandings of history from the popular consciousness, but also depriving the masses of the discursive tools to appreciate how their material denigration and social vulnerability related to the historical victory of "market rationality". Instead, their means of formulating an understanding of social reality were limited to an incontestable set of assumptions around 'identity politics', 'xenophobia', or '"the culture wars"' strategically positioned at the extreme centre of the so-called cultural imagination (Khalil). Forced to the periphery of understanding were acute economic crises, social structures in perpetual collapse, and states subject to political humiliation and economic stagnation; all of which were exponentially more meaningful in their implication as sources of future instability.

In the Arab world, the collapse of the Ottoman Empire and its bellicose division by its Christian imperial rivals marked the dissolution of the caliphate, provoking a violent disorientation of its existing social structures formerly based upon a common language, history, and culture. The nihilism of the neoliberal paradigm further threatened the caliphate as a structural point of reference, as the now socially fragmented Arab states marched blindly into becoming free-trade industrial zones, their populations made causality of new type of imperial war built upon the struggle to wrest international capital from one's rivals. Within the context of this 'perverse world, stripped of the discursive space necessary to articulate any opposition to neoliberal ideology... al Assad can successfully market himself as the civilized man in a battle with brutes from 1000 AD while killing hundreds of thousands of Syrians' (Khalil). Similarly, the extremism of the Islamic State can be looked upon as a co-production 'of the Western and Muslim worlds, specifically of Western and Muslim "security" agencies who have lost all moral compass in the pursuit of geopolitical prowess, self-aggrandisement and corporate profiteering" (Ahmed). The neoliberal economic restructuring of societies have made for a profound sense of social dislocation, but also temporal distortion, where analogue visions of imperialist warfare, capitalist modernization, and technological determinism digitally accelerate to stage an image of a portentous future in the so-called cultural imagination.

The Islamic State's social media strategy relies fundamentally on the amplification of 'a sense of temporal acceleration and imminent arrival of end-times' amongst its supporters (Berger). Their approach works by 'leveraging the dynamics of social contagion and remote intimacy' already assimilated into the everyday lives of the postmillennial generation under conditions of neoliberal hegemony. Its highly sophisticated media outputs are designed to provide its followers with a heightened sensory experience, affording them the opportunity to conceive of life under an alternative form of sovereignty. It is specifically designed 'for supporters outside its territories to immerse themselves in a highly idealized version of its millenarian project, the so-called caliphate' (Berger). The Islamic State is adept at exploiting the data availability and relative transparency of social media platforms, and has become notorious in its 'bombing' of Twitter's structure, 'hijacking' the most popular hashtags of the week and using them in IS-related posts. Their supporters have been known to simultaneously post thousands of tweets featuring the same hashtag, and repeat this throughout the day in order to generate a larger online presence far greater than their actual number of followers. Such actions subvert overall activity in the network, allowing IS to temporarily dictate its motion.

In addition to this, IS have been able to produce their own external media channels in the form of communiques and news broadcasts. Their stock imagery consisting of military footage and civil society content in roughly equal proportion is intended to reinforce a perception of their legitimate statehood. IS's civil society presents itself visually as the ultimate society of control; populated with police officers, traffic controllers and public executioners, poised to rapidly capture and destroy anyone engaging in acts of deviation from the ideological norm. IS has borrowed heavily from Nazism to build up its state branding that extends far further than the adoption of its own state salute. The hyper-prominent display of its flag, emotive imagery of apocryphal belonging, emphasis on ecstatic martyrdom, character as an invincible fighting force carrying out a cosmically destined mission, and privileging of filmic propaganda to mirror the advanced production values of a greater ideological competitor, i.e. Hollywood, all point towards an emerging form of statecraft where a medieval absolutism can be fused with corporate militarism in order to encode territorial domination with mythical implication.

Within this scenario technological warfare plays out in two parallel, interrelated worlds: the combat zone and the digital device. The desire for warfare amongst the two is heightened by their encoding of their surrounding states and ideologies with reference to past historical contests that bear little obvious connection to the postmillennial age. Its youngest generation has no developed sense of history, geography, language, or religion to appreciate the origins of their frustration. As a consequence, they prefer instead to withdraw into virtual landscapes where other forms of validation and justification are made seemingly possible through the formulation of a new identity and the adoption of a political vision that promises to bring about an inverting of the social order. 'The majority of

French Muslims going to Syria are now middle-class youths, some of them white converts to Islam, and an increasing percentage of them female' (Packer). Their rapid radicalization comes by way not of their families or communities, but primarily through social media imagery that compels them to emotionally identify with the violence perpetuated against their fellow-Muslims around the world. Once they are inside Syria and confronted with the reality of the Islamic State's violence on the ground, 'some are shocked … and try to return home; others are seduced by it' (Packer). Whatever happens there can be no political reconciliation of the geographic, social, and ethnic grievance felt amongst those inhabiting these mutually occupied territories of binary apperception.

Those who find themselves captured within the milieu of 'no culture' find themselves the most vulnerable within this apprehensive perceptual economy. It is they who desire to exert their revenge upon the apparatus 'not only by asserting their humanity (or what appears to them as such) ... but by placing that apparatus in the service of [their] triumph' (Benjamin "Art in the Age Second Version" 111). If film expends its cultural capital for the purpose of countering revolution, there can be no political advantage drawn from it that might present an opportunity to put an end to its powers of exploitation. Rather what must happen is something entirely different in the way the apparatus is approached not as something that offers greater self -understanding, but rather something that gives a crucial insight into the ways in which humanity is being aggressively moblised to reproduce a society of control through our fascination with the capture and differentiation of ourselves and others, the endgame of which is always war.

Ending Times

The contemporary world is governed by speed, a "dromoeconomy" governed by a combination of war without end and an overproduction of trade. Together they function as post-state actors to 'command, control, solicit, and deploy highly sophisticated information technologies, including, and especially, communication technologies' (Armitage and Graham 113). Within this schema, 'consumption facilitates the politics of resource and territorial management; technology controls communication and transmission of energy at its base forms into the complexities of every facet of life' (Colman 246). Living in a dromoeconomy means negotiating a material field characterised by an excess of speed and power that is dependent upon surplus time, surplus value and surplus labour. At present the postmillennial generation dwells within an 'overpopulated, asymmetrically resourced global system', what Paul Virilio describes as a "grey ecology"' (255). The finiteness of this world has become its own obsessive object of contemplation 'played out multiple times in films of apocalypse, in video games, and in other screen-based media, where one of the most popular genres is horror, a horror of the world' (255). A generation who understands that it exists in what Felicity J. Colman terms a 'revelationary post-world-end state' must have the meaning of its presence directed elsewhere and its unreadiness to confront its reality mediated towards various forms of distraction (255).

One way to read this 'revelationary post-world-end state' is through what Alexander R. Galloway refers to as 'the control allegory' (xi). Within this economy the screen becomes the site for life itself to become synonymous with continuous labour. For it to display its valour it must be performed in a privileged state of distraction while performing banal life tasks such as driving a car, walking down a street, or putting a child to bed. In each instance, the individual is informally tethered to their digital devices, and obedient to the call of being in a constant mode of production. The great irony is 'each and every day, anyone plugged into a network is performing hour after hour of unpaid micro labor' and in so doing gradually acclimates themselves to unremunerated labour as a normative condition (136). Life is being subtracted from itself on a microlevel and in a way not dissimilar to the menace of capture that film once exclusively portended. As the world's first new media, Lev Manovich argues film 'was always already a digital' format (4). Film as a technique of mechanical automation, lent a verisimilitude to a totalitarian ideology of 'essential purity in which only certain subjects portrayed as dominant,' and all others consigned to an inveterate 'alterity' (142). Digital media now serves to legitimise and promote a 'customized micropolitics of identity management, in which each human soul is captured and reproduced as an autonomous individual bearing affects and identities' (142-143).

Film's digitisation has made it no less an apparatus for reinforcing this order, through its assignation of certain identifiable traits to distinct classes within society and manufacture of complex narratives for explaining and maintaining such hierarchies as an inborn phenomenon. 'Ethnic and racial coding seem always to be synonymous with mediation itself' and the fate of bodies determined by the way filmic apparatus compels them to perform restively within a certain affective identity's limited contour (Galloway 133). The passive reception of what is being projected cannot be classified as a leisure activity, or a frivolous encounter so long as the masses continue to perform the productive activity of making operative a 'system of subjective predication' (144). In this sense, the screen becomes a forcing arena for the digital repression of certain classes.

Within this scenario the tools left open to them, as Benjamin predicted, are interruption, illogic and friction to break with the precarious durational order by which they have become enveloped. The illusion that there is no outside to this order can be shattered to expose the contingency, fragmentation and ruin upon which it rests, making visible the violence and death that are the true cost of its production. The device that Benjamin believed could achieve this unveiling was allegory. Allegory possesses a structural complicity with what it represents and therefore is capable of a dynamic mirroring and a criticality toward what it reflects. Its apparent complicity with the instrumentalising economies of new media makes it the perfect foil for their destructive force. 'Allegory oscillates between a project of reference and a project of capture' and therefore comprises the perfect positionality from which to critique the present political economy (Teskey 8). Benjamin observes that 'allegories fill out and deny the void in which they are represented (Benjamin 227). Allegory therefore is redemptive, because its structure is made up of both violence and counterviolence and in this way 'its destructive impulse retains the very state that it both contains and contests' (Sanyal 163).

The world does not appear in our image, but rather is itself a reproduction of 'illusory displays and ambiguous speculations' that dissociate and dissuade humanity from realising new social opportunities that are only recognisable when the desire for fragmentation is abandoned (Benjamin "Art in the Age Second Version" 114). Until then, digital technologies continue to provide 'additional possibilities for the creative wrecking and degradation of almost anything. They multiply options for destruction, corruption, and debasement through 'digitalized warfare, the computerization of production, and real-estate speculation,' which make abundant capitalistic artifices with a tangible physical component (Steyerl "Digital Debris" 71). Here allegory can be called upon to reveal flaws in what appear to be the uniform background of these reproductions and start to lift away the apparatus to inspire a present understanding of ourselves that does not rely on 'being torn between mourning past demise and a violently displaced future,' nor on following the orders of menu-driven identities (71). Here is it possible to surmise that Benjamin's concept of history is really about its progressive voidance under modern conditions of knowledge and power. In Benjamin's time, controlling the means of production was the preferred way to monopolise power. Today this happens through the control of information. The digital infrastructure of the twenty-first century makes it possible for power to operate through a mathematical means, and for factual material to be manipulated in the interest of multiplying the means of control. Within the frame of that distortion, everyone can see the future, but no one remembers the past. Allegory responds by distorting that distortion.

Shock Tactics

In the early part of the twentieth century, modern forms of information and communication find their audiences in the conveyance of shocking events or experiences. Benjamin's work prompts us to consider why it might be that the theme of shock so prevalent in cultural understandings of that period were largely abandoned at the end of World War II, that is to say 'at the very moment … when information and communication were naturalized as the accepted form of knowledge and language throughout public life' (Day 94). In 1939, in his essay, "Some Motifs on Baudelaire," 'Benjamin laments how the modern subject's consciousness must be enhanced – through a psychical mechanism of neutralization – against the overwhelming, abrasive shocks of everyday life' (Hall 89). Benjamin believed that the casualty of efforts to preserve humanity's sanity was the loss of his ability 'to assimilate and voluntarily recall information' (89).

Benjamin was wary of the visual decoding of knowledge, and sought to recover meaning through another sensory means: sound, or more specifically, noise. Benjamin recognised that 'subversive noise' was the object of greatest threat to a looming totalitarian state, which recognised within it 'demands for cultural autonomy, support of differences, or marginality' (Attali 7). Its conduits functioned from a fundamental refusal of abnormality, a need to illicit confession, and a desire to listen in order to silence certain voices. 'The dreams of political sciences and fantasies of men in power' chime at the point of recording, probing and memorising certain patterns so that they might 'interpret and control history', 'manipulate the culture of a people' and 'channel their hopes and violence' into a tool of cultural repression. i.e., 'noise control' (Attali 7).

Benjamin recognised that historiography could act as a revolutionary system of organisation based on its capacity to reveal humanity's historical record as a phenomenon steeped in 'appropriation, pulsing, overlapping, phase-shifting, resonance and feedback' (Hight 15). Once acknowledged as such, it can be made otherwise operational to dissemble authoritarian forms of power at their molecular level, so they can no longer rely on the continuity of habitual, consensual perception to achieve their ends. Benjamin's employment of sound as a tool of cognitive breakage, acts as a means of calling out the invisible tyranny of homogeneity, and indeed its relationship to technology as an occupying force, one that denies agency to one's experiences within the span of modernity. Benjamin's work then must be conducted at the subtle frequencies at which technical gaps emerge, and it is possible to engage otherwise the instruments of reproduction and recording and their actions upon the subject's very existence.

In the third version of the 'Artwork' essay, Benjamin refers specifically to Freud's *Psychopathology of Everyday Life*, as having 'isolated and made analyzable things which had previously floated unnoticed on the broad stream of perception' ("Art in the Age Third

Version", 265). From this vantage point Benjamin is able to concede that, 'a similar deepening of apperception throughout the entire spectrum of optical - and now also auditory - impressions has been accomplished by film' (265). This 1938 version makes new reference to acoustics, and indeed, 'seems to suggest that the whole breadth of the acoustical realm is now also receptive to heightened sensitivity, at least through film' (Ryder 139). Through a similar means the acoustic opens a gateway to the unconscious through noise that is capable of both awakening and disrupting impressions made to the mind. For such noise to ultimately convey meaning, a modern set of 'ears are needed, that are attuned not only to 'the retreat of sound' but equally to its arrival (Richter 195). According to Gerhard Richter, such ears 'must register what is not fully itself within that which penetrates them, what remains other and elusive in the hidden figures of the acoustic. To say that the ear must be more than one and that it must remain open to its own plurality and that of others, is to begin to think through the ways in which it is inhabited not only by the self … but also the voices of others' (195). Such co-habitation between oneself, the Other, and new technological apparatuses of acoustic projection naturally lend themselves to a recognition of the obscured hybridity of that co-voicing. Beyond that, ears that are technologically attuned might provide a means through which to disclose the subjective ambiguities reverberating beneath the surface of everyday life.

Benjamin's fascination for the political potential of this plurality was at direct odds with the authoritarian character of politics that coincided with modernity, which constantly experimented with new and lasting technologically mediated forms of social control. Included amongst those formats where the telephone, the gramophone and the radio. Each was premised on masking a coercion from without as a coercion from within, therefore, leaving the subject utterly bereft of the psychic means to exert control over the technological apparatus. The voice of authority must transmit itself as a singularity through repetition so that it may bypass any threat of a heterogeneity of interpretation. This voice is nonetheless haunted by the echoes of alterity and the potential of listening as a politically resonant act.

Benjamin's recourse to lived experience, a political impetus profoundly rooted in purpose, and the apprehension of a duality between reality and consciousness, meant that his critical project would not only anticipate the emergence of new technical systems such as photography, film, and sound recording, but would inadvertently contribute to the technologising of perception he primarily sought to expose. Benjamin's work was devoted in many respects to understanding a world that was to grow exponentially smaller as an effect of the profound traumas of financial crisis and total warfare. Social science was taking the place of philosophy as the omniscient interpreter of the meaning of perception and memory. As a consequence, the motivations of its subjects were placed under unbounded scrutiny, while at the same time, humanity's conception of itself was carefully limited to the categories of psychic internalisation, self-realisation and mediated thought. Over time what would emerge was a self-referential world, which was fundamentally relational.

Visceral Disordering

In 1932, Benjamin pursued one significant avenue through which to depart from a strictly visual decoding of sensory knowledge. During a 'field' visit to Marseilles, Benjamin conducts an experiment on himself to record the effects of consuming hashish. His preliminary remarks about his experiment address his sense of phenomenal alienation subsequent to his ingesting the drug. He recalls how 'his laughter, his utterances, happen to him like external events' and how 'in the middle of a sentence these transitions can take place' ("Hashish in Marseilles" 672). These observations place Benjamin in a rare position of externalised subjectivity, where these happenings arrive to his conscious mind 'in a form that ... diverges very widely from the norm' (672). These deviations in his self-understanding emerge when 'connections become difficult to perceive, owing to the frequent, sudden rupture of all memory of past events (672). What is remarkable is that an awareness of such breakage 'is not formed into words' but rather through compulsive laughter and that he as a temporally dissociated subject is capable of 'nothing except laughing' (672).

Laughter, 'in Walter Benjamin's words, is "shattered articulation"' ("Those in Glass Houses"). Esther Leslie describes laughter as having the capacity to 'break[s] up both words and the body, and to disarticulate, 'everything' that might surround those forms ("Those in Glass Houses"). What is most notable about Leslie's account of laughter is that it acts as both a literal and figurative disruptor: 'A person in movement might be stopped in their tracks. A person speaking has the stream of words cut off. The listener hears only a clatter of stuttering sounds. Laughter is an interruption to the ongoingness of life and meaning' ("Those in Glass Houses"). Laughter's power to disorder the flow of movement and expression is something that has the capacity to dislocate events. It distends time, and lends it a looping character in which events take on the sensation of a vibration, and become conversant with laughter as an involuntary explosion of noise and movement beyond the body's normative rhythms.

This highly resonant encounter with shattered forms of articulation, continues to play out in this same essay, when Benjamin refers to the sound which animates his liminal experiment. Benjamin describes it as 'persistently rising and falling; "rush[ing]" and switch[ing]' (678). The music he is referring to is improvisational jazz. Benjamin finds jazz has several unintended effects upon his body. He finds that he has forgotten 'on what grounds [he] permitted [himself] to mark the beat with [his] foot' (678). He observes that such an impulse' is against [his] education, and as a consequence, his actions 'did not happen without inner disputation' (678). Finally, he admits, 'there were times when the intensity of the acoustic impressions blotted out all others' (678).

Benjamin experiences jazz as something capable of subverting his bourgeois upbringing. Within the confines of his own body, he recognises that a type of sensory retraining is subtly taking place, capable of lending his mind an ultra-heightened form of impression. Similar to his experience of laughter, Benjamin finds that as normal bodily control cedes to involuntary impulse he no longer desires to exert control of jazz's effects upon him. The intensity of both phenomenon present Benjamin with a sensation that the coherence of both movement and events are illusory, and beyond their surface may lay intensities his body has yet to assimilate.

Through the recapitulation of both experiments, Benjamin makes evident the revelatory potential of parting ways with social convention so that one may reveal to themselves the more arcane dynamics at work in the shaping of human perception.

The improvisational jazz that Benjamin was audience to in Marseilles, relied upon summoning the effects of alienation and directing them back towards jazz's domain commercial forms captured through sound recording. Through such a relay of disruption, it sharpened the attention of active listeners like Benjamin. Howard Eliand observes that improvisational jazz required the practise of a 'high-speed vigilance,' which like its listeners, was 'conditioned…by the dynamics of modern technology (56). The reprogramming of the human body to relay itself through the modern industrial cycles of acceleration, transition, and obsolescence prepared them to engage with things such as modern media technologies and mass-produced commodities as though they were mere extensions of their own natures. Improvisational jazz interfered with this process of assimilation by decoding and remastering the technological apparatus humanity was preparing to readily adopt as its second skin.

Improvisational jazz, while remaining integrated with the apparatus itself, was able to nevertheless raise humanity's consciousness of the 'interpenetration of nature and technology, physis and technē' making its listeners aware that their basic survival within the context of modernity, depended on their 'creative powers of absorption' (Eliand 56). As a kind of skilled technician, the improvisational jazz musician, 'must have at his disposal a set of (variable) moves, to paraphrase Benjamin, in order to perform this task, which involves equal measures of spontaneity and knowledge, or receptivity and productivity. The deflection of attention here is manifold and concentrated, for the player is both carried away and in control' (Eliand, footnote 10, 65). The technique that the improvisational jazz musician used was to overwhelm his listener through the intensity of the acoustic impressions he laid upon them as a vital part of achieving his decisive operation. Essentially, his aim was to shock his listeners.

Shock presents the receiver with an instant of insensibility 'as the primary experience of dislocation in modern life and as an aesthetic practice to free art from the enslaving and exploitative dynamics of commodity capitalism' (Ezcurra). Benjamin saw 'shock as a key to changing the mode of reception of art and to disrupting the dismal and catastrophic continuity of everyday life' (Ezcurra). Shock for Benjamin was a way to deal with the excessive violence imposed by the conditions of modern life, a way to apply a sensory break so that stimulation could be assimilated into consciousness as information. Benjamin foresaw the use of shock as having two potential outcomes in terms of forming a new pattern of social reality. Shock could be utilised as a means of 'shattering the conformist, blinded man who lives in a state of alienation, and make him so uncomfortably "strange" that his curiosity may be aroused' (Ezcurra). Alternatively, it could be 'employed to promote the emergence of a mass of "traumatized automatons;" the vanishing of private space; the coming of an era where experience, devoid of tradition, is incapable of finding meaning'(Ezcurra). Benjamin's great hope for the future was that mass perception would oscillate between extreme and apparently irreconcilable positions. So long as this duality persisted, it provided humanity with a means of expanding its own freedom.

Improvisational jazz, like the experimental forms of filmic montage that captivated Benjamin's critical imagination, was 'a creation born of modernity and thus its advent mirrors the 'contradictions of Western industrial capitalism' (Kodat 4). While these emergent media forms seem on one level 'to speak the language of resistance and subversion', at another they remain 'enmeshed in the imperatives of capitalist production and consumption' (Kodat 5).

Weimar era German listeners found pleasure in jazz based on its shock value, that is to say its ability to confound and confuse the sensibilities of its listeners, and to attune them to noise, as a source of both meaning and desire. It taught them to desire an unintelligibility that was a product not of high culture, but rather, street culture. This music therefore acted as code that was from on high, indecipherable to Germany's cultural betters. Its enigma appealed to a much more rudimentary logic, insofar as 'the music seemed to enter into and possess the body' (Wipplinger 311). Goebbels referred to jazz as the art of the subhuman, as a medium that found its currency in proximity and contact, through muscle and reflex.

Such music reveals to the listener the noises of life and the body that are monstrous, marginal, and obscure and allows them to take pleasure in this anarchic sound, and for to them to play and to improvise once again in the key of their own being. Improvisational jazz allowed black power music and European abstract, theoretical music to co-mingle and in doing so, broke down their repetitive hierarchy, inserting in its place new codes upon which to construct a new culture. 'In the slang of the black ghettos, "to jazz" and "to rock" both meant to make love' (Attali 103). The ultimate counter force to institutional violence and the circuitry of capitalism ultimately came down to love, to making love amidst the reality that life itself is a risk, and that the struggle against and endurance of violence is the very thing that gives life meaning. Improvisational jazz breaks completely with caution and with the repetition of a history that demanded black alienation, containment, repression, limitation, censorship, and expulsion. Nonetheless, within the greater society these persists as the conditions that organised reality of what black bodies could do and where its expression belonged.

Arresting Jazz

For many critics, including Benjamin's contemporary Theodor Adorno, jazz wasn't music at all, insofar as it lacked composition and refused to conform to 'predictive reality' (Attali 147). Adorno wrote in a letter to Benjamin dated 18 March 1936, that he was about to pass 'verdict' on jazz 'by revealing its "progressive" elements (semblance of montage, collective work, primacy of reproduction over production), as facades for something truly quite reactionary' ("Letters to Walter Benjamin" 125). Adorno explains further, that he believes he has decisively 'succeeded in decoding jazz and defining its social function' (126). Adorno's criticism of jazz strongly resonates with his observations in that same letter concerning 'the laughter of audiences at the cinema' … as 'anything but good or revolutionary; instead it is full of the worst bourgeois sadism' (123). In both instances, Adorno seemed to be convinced that commercial distraction of any kind from the rigours of capitalist labour entails conformity with its inherent reactionary potential. It does nothing to progress society, but rather presents its audiences with nothing greater than a dangerous release of pleasure intercut with cruelty. Therefore, such forms are inherently crude and thus counter-productive to the cause of social revolution. Benjamin's critical premises to the contrary were deemed by Adorno to be either hopelessly romantic or politically deluded. At the close of his letter, Adorno, informs Benjamin in no uncertain terms that he would simply have to shift into line if his arguments were to be taken seriously by his Frankfurt School peers.

Adorno first wrote disparagingly of jazz during the initial years of the Third Reich. It is significant to note that during that era, musical composition was overwhelmingly the domain of academic institutions as opposed to the entertainment or recording industries. The great irony is that while Schoenberg, Adorno's teacher, himself produced 'relentless serial noise as a protest against the complacent bourgeoisie', Adorno pinned his hopes for the future on being a musicological critic radically opposed to the acceptance of atonal music (Attali 153). At that time, Adorno found himself in the process of losing his teaching privileges at the university. He made the strategic decision to write a criticism of jazz that was 'complimentary to the National Socialist political establishment's condemnation' of it, essentially as a means of remaining within the academy (Kater 14).

While Adorno struggled to stay in place within the existing bourgeois order, a radically displaced Benjamin sought to situate himself within the historical force field that was Nazi Germany. In so doing, he radically exposed himself to the fore-history and after-history that was being played out in National Socialism's crystallization as a politic. What Benjamin set out to do was to shatter the apparent continuity of its relationship to an earlier mythic form of Teutonic power and knowledge and to introduce in its place a cognitive echo of the cultural montage that had been the true phenomenon supporting the modern German state's founding.

Such construction must always be premised on deconstruction. In noise, Benjamin found the acoustic technology fit for his programme of resistance. Noise was capable of restoring human perception back into the equation of the technological exteriorization of cultural and historical memory. When repositioned in this way their illusory character of unity and linearity could be exposed as the false origin of their truth. What is subsumed by this historical logic can never be destroyed, but will echo forth into the afterlife beyond its inevitable ruination. Eventually it will emerge as a pulse that will signal what is to come for those with the polyvalent sensibility to hear it.

In the meantime, Benjamin would use his extraordinary sensitivity to noise to negotiate his way along the teleological circuit that was National Socialism, and probe along the contours of its collective technological psychosis for a necessary means to survive German 'civilisation' if need be.

Benjamin was not alone in this endeavour to seek an expressive form of societal redemption. Leftist Jews in Nazi Germany would be drawn to jazz because of its expression of free will, and its resistance to imperialism. In racial terms, its origins and dissemination were pronouncedly non-Aryan. Its alien, syncopated rhythms lent themselves to love not war; their complexity overwhelming to any simplistic message of order.

German Jews were admiring of (African) America's deep inroads into cultural life in modern Germany, and appreciated the power of its alterity to articulate itself amidst a historical geography of racial segregation. German Jews felt an affinity with an African America compelled to situate itself relationally to a white supremacist American nation. As a sub-culture, it too had to find a way to simultaneously pronounce itself as an ontology, and express itself as an epistemology. This identity had to perform its authenticity by adopting a position that would resonate with truth at a different level to what was projected upon it. What German Jews perhaps admired most about African Americans, was this subgroup's ability to convey difference back to the totality of the American social experience persistently decreed as "white", and imbuing it with a deeper sense of itself as essentially the product of montage.

'Benjamin once described philosophy as a "harmonious relationship"' (Hall 91). Race, insofar as it functions as an analogy for organisation and identity, exists as 'a category built into the very foundations of Western metaphysical and epistemological systems' (Hight 15). Benjamin recognised harmony its the ultimate measuring device.

Harmony acted as a means of totalling up bodies and subtracting from them a degree of likeness, wherein they could only be acknowledged to be assimilated and controlled in relation to whiteness. Anyone appearing against this measure was marked by deviation.

An offbeat-ness of alterity, lends figures like Benjamin, a sense that their lives operate on a slightly different register of time. Such time compels them to either move ahead or fall behind, to progress imperceptibly within the branching, intersecting pathways of a nodal time whose points within the diagram of whiteness coerce one to stand still, or leap ahead. Within such a time span it is possible to slip the breaks, and observe at close range whiteness's purportedly homogeneous space. Here one can dwell within its logic of sameness, and deftly navigate its historical composition of race 'interleaved into an interstitial abjection' (Hight 15). It is possible to see race itself from a new perspective through acceding to this 'temporal occupation', and reveal to oneself 'that the white grid's natural harmonies have been, in fact, constructed' (15).

Time is the register on which this danger of expulsion can be transformed into a sense of cogent awareness. This takes place not through offering an alternative system of emancipation, but rather by reconfiguring the tone in such a way that it does not merely invert the harmonic system towards a new standard bearer of universalism. Rather, it is where the subject gains the radical sense of a sonic landscape of power that has existed all along.

If shock value is concentrated and accumulated, located and situated at certain figurative locations, the question for Benjamin becomes about how to get to it, and how to temporarily break there as a means of intruding upon it and installing parallel structures. Benjamin's work was devoted to uncovering worlds that enabled us to initiate a different type of stake within this existing system of values, and therein dodge the apparent totality of this particular scheme of reality. Benjamin's work asked us to consider how humanity became joined to this system of value. Equally, it experimented with how humanity might locate itself in a way that at once allowed it to tactically survey its outlays, and misapply its position so that these world-making practices might be revealed to it.

Noise Generation

The Cold War world of the late 1940s and 1950s was populated by a generation of 'survivors' whose paranoid perception of the world was profoundly influenced by the horrors of the Great Depression, German fascism and the advent of the nuclear bomb. Their lingering existential anxiety would rapidly translate into a kind of mania for procreation in Europe and North America during the 'baby boom' years of 1946-1964. What had been known as the 'Silent Generation', because of their aversion to deviating from social norms, would eventually give rise to the youth culture of the 1950s and activist culture of the 1960s.

What American teens would have in common with their European counterparts was a cognisance that far from disappearing from the Western psyche in 1945 a kind of crypto-fascism had in fact seeped into every aspect of postwar civilian life. As the United States emerged from World War II as the world's latest cultural hegemon, it was those industrialists who profited spectacularly during the Nazi-era, such as IBM's Thomas J. Watson that now seamlessly lead on the development of America's postwar economy.

As the Cold War emerged as the newest theatre of total war, former Nazis and Nazi collaborators were recruited to work in every corporate, political, and educational institution - including the police and the military - in both Germany and the United States. It is at this point of historical interface between capitalism and authoritarianism that cybernetics comes into its maturity, promising an unprecedented co-joining of the mind with matter, body with consciousness, reality with representation, and philosophy with engineering, as a means of producing a predictive universe in place of a probable world.

Cybernetics introduced a second nature into the postwar consciousness of an emerging youth culture. It projected the techniques of sensory stimulation that worked on young American white boys in the 1950s onto an entire planet. Cybernetics rewired this generation's imagination into intelligible patterns of communicative feedback. Cybernetic signalling was the new 'rock and roll', a device that allowed wakefulness and reverie, command and solicitation to course over the bodies of its affectively engaged listeners. These bodies were being attuned to new forms of ordering power, knowledge, and culture, making them intelligible across all previous boundaries held between man and machine. Information became a universal currency to unite their features. Information became accountable for them, projecting them into the future, guaranteeing their return, promising them security, options, and futures in exchange for their willingness to supply their energy, feedback, and conversion to enliven this nodal exchange.

Within a system of cybernetics extra credit was assigned to those carrying out a masculine, heterosexist, racially inscribed charge. Cybernetics implied the control of some subjects over others. It was designed to subtly promote the further destabilisation of certain proscribed others. It forced them to adhere to a teleology that sacrificed them up to a new God of technical mediation. In this case, God was now put in charge of an agency comprised of letters, icons, and moving pictures. Whereas previously he had been a God of representation, in this new schema he emerged as a God of repetition. God was expected to time his labours to coincide with reflexive power and to rewire his older demand for sacrifice into these newer protocols of control. What started as the word, became the body, became energy, became labour, became information, became capital - the ultramodern form of authority.

All is not lost however for the potential inherent to Benjamin's prewar revolutionary programme. What stands in the way of cybernetic omniscience is noise. Noise, identified here as an overload of information, would make universal predication impossible as an effective target. Noise of this nature would have to be co-opted and operationalised as it is extraneous to the surplus value of information. The threat of noise is that it cannot be translated readily into meaning. Noise implies tension, a refrain from the instrumentalisation of sound itself. Noise resists the locatability, commensurability, and translatability of a sovereign perceptual field.

By mid-century, this field had already been profoundly compromised in its authority through the advent of industrial capitalism, imperial warfare, human holocaust, and nuclear annihilation. In the desire to achieve economic homogeneity and technological convergence humanity has neglected Benjamin's call for an excavation of these forms, and for the alarm to continue to sound from these events within the duration of memory. This poses the question of how we can conceive of an 'atmospherics of noise' that may destabilise and disarm authoritarian political apparatuses in the present.

Walter Benjamin's work toward the end of his life explored the critical differences between surviving and living on. This difference can be understood as an engineering problem, something that can be said to rely on a logic of calculation versus articulation. Benjamin's recourse to noise as a fundamental terrain of incompatibility, meant that as an object it fundamentally stood in the way of transmission, and therefore acted in a way that was discursively irreconcilable with the aims of both information and communication from a nascent cybernetic point of view.

What Benjamin ultimately endeavoured to refuse was the one-way street of rationalisation within the system of meaning. What he dreaded most was for leftist ideology to become crystallized in the figure of capitalist realism, so that it became a popular target for fascistic violence. His work thus signalled that another genealogical line of flight is possible, should one be willing to assimilate the anarchic, accidental, conjectural, and contingent into one's conception of the future.

For Benjamin, to break with functioning along the lines of an informational order accelerated the disruption of the ideological, political constitution of reality that subtends its founding. In our current digital age, we would do well to refuse a promised transparency amongst 'information systems, humans, machines and nature' (Pinto and Franke 27). Indeed, we must fight to preserve an atmosphere where it is still possible to detect the shockwaves that travel between them, carrying a resonant signature of violence and expropriation. Universal connectivity as a concept carries with it the charge of coercive participation within a hyper mediated milieu of being. Such totality at once codes and coincides with authoritarianism and eventuates, in time, a need on one level for a limitless expansion of power, and on another for a modality of profound exclusion. It becomes impossible to distinguish where one level ends and the other begins. 'Hence the need to perform one's complicity' in hopes that it will prevent one 'from falling into the ranks of surplus populations that are rendered invisible, voiceless and ultimately non-existent' (Pinto and Franke 29).

In the beginning decades of the twenty-first century, humanity once again finds itself being primed for mobilisation through a modality of engagement that will train it for a total war to come. This war will be fought to defend capitalism against democracy and guarantee rights only to those who can claim a digital nativism. Within such a precarious economic terrain life is the exception to the rule of death. The way to counter such a threat for Benjamin is never to conform to its standards. Instead, humankind must to double down on death itself, and to introduce into the equation, 'a consequent second life, a survival of life not understood as its continuation beyond a terminus within a common, objective frame, but as its reemergence at a new location in history with the fully eruptive force of its original appearance' (McFarland 210).

Benjamin's philosophy, therefore, is inviting a meltdown of time and space itself, prompting a fissure in the continuum of life and survival, such that theirs is exposed as an arrangement of mutual dependency. This is not an act of mimicry, it is a gamble with mortal displacement, a faith in the afterlife – of meaning. In the end, Benjamin wanted us all to become translators (as opposed to venture capitalists and hackers who venerate fantasies of technological liberation), so that we continued to rely on our own hearing, forcing ourselves firstly to sound off before assuming we were understood.

Similarly, those with revolutionary capabilities had to first fail their auditions, and in this way, forestall the arrival of final judgement upon their life's work in the act of heralding the end of all times. Benjamin implores us to stay in the realm of acoustics so that we can use that as a means of inserting ourselves into oblivion through the looping conduit of history. In terms of articulation, 'one ought to speak of events that reach us like an echo awakened by a call, a sound that seems to have been heard somewhere in the darkness of a past life. Accordingly, if we are not mistaken, the shock with which moments enter consciousness as if already lived usually strikes us in the form of a sound' (Benjamin, "Berlin Chronicle," 634).

Time in this sense is uncontrollably spreading, delivered in haste, by a 'punk messenger', 'who set in motion is a pointer, he blinks in the direction of writing, which carries war and other cries, his screams are being dictaphoned by the inserted cries of murder and pain, spreading everywhere' (Ronell 63-64). We too find ourselves standing on end, the product of an evolutionary aesthetic that demands that we split into our 'dividual' form, mimicking a path of nuclear fission and yet somehow, we sense the future is over. Perhaps what remains is simply a tantalising prospect of being included within a digital afterlife; a place where there are no gaps, and everything is automatic.

This desire for programmatic erasure becomes the stuff that will power us through the decades of the 50s, 60s, 70s and 80s that commence a negation of negation. These decades are characterised by a refusal to take seriously the outspread of despondency and deep-seated resignation that characterised Benjamin's age. Over time they herald into place a cord of optimisation 'that conflates the vectors of Silicon Valley commodity space with the strategic space of the United States empire' (Pinto and Franke 29). As a consequence, humanity tilts towards a fate of endless repetition, so that history itself becomes a loop stranded within the hermeneutic casing of American power. As the Cold War ensues much of what Benjamin picked up as fascist violence will be subsumed into the measures of what we refer to as the 'smart state'. This makes it difficult, though not impossible to detect that the fusion of cybernetic technology and digital reproduction, remain at some level fuelled by delusions of racial superiority and genocidal extinction. This subtle reverberation will make it possible not only to receive the orders of a technical determinism, but also paradoxically to overcome them.

Chapter 3
Black Mirror:
Binary Apperception and Digital Racism

Over the past five centuries race has played handmaiden to economics. An evolving concept of race prepared the ground for the arrival of capitalism, which thereafter originated 'primitive accumulation, worldwide European seaborne empire, the Westphalian state system, conquest and settlement, the African slave trade, and the advent of enlightenment culture' (Winant 1). Racism in its present form is a specific product born of that history; its physical markers crafted to support these various systems of human hierarchy. Race manifested itself firstly as a cultural abstraction that aimed to capture individuals at the level of desire and comportment and install within them a nexus of traits and attitudes that corresponded well to the dictates of commercial empire. The cultural coordinates of race charted European civil society towards the expression of gentle manners and rational intellect, and refined its sensibilities to include an acceptance of the violent subordination of peoples abroad whom it was believed were not able themselves to express such attributes. Eventually skin, muscle, blood, and bone will be brought into the equation to reify racial classification, and in so doing, permanently implicate it as a feature determined in nature, rather than culture. What had begun as a judgment of moral and intellectual abilities progressively morphed into distinctions in physical characteristic, such that race became something reducible to organic material and civilisation the raw ingredient for planetary domination.

Empire needed racial subjugation to appear as inexorable, and for race to slowly conflate itself categorically into racism over the long course of its social, political and economic evolution. As empire was expanding abroad, its racialised articulation was insinuating itself into the foundations of modern political and civil institutions at home. Racial essentialism was never meant to be merely skin deep, but was designed to penetrate deeper into society to rationalise commercial inequalities. The complex assumptions built up around race, therefore cannot be easily separated from a greater history of commercial governing. In this sense, racial ideology was always intricately tied to concerns around trade and profit, with those making their fortunes involved in the enslavement of African men, women, and children on plantations in the Americas and West Indies acting deliberately to promulgate an acceptance of the necessity of racial subjugation. Their commercial need required that race move from a space of geographical location, to one of categorical confinement.

Human Cargo

The earliest known incidents of slave ownership in Britain took place from the 1570s onwards, wherein Peter Fryer maintains they were used in three main capacities: 'as household servants (the majority), as prostitutes or sexual conveniences for well-to-do Englishmen and Dutchmen; and as court entertainers' (Fryer 8). Towards the end of the sixteenth century it became fashionable for the landed gentry 'to have a black slave or two among the household servants' (8). Amongst the first acquire one was Lady Raleigh, the wife of Sir Walter Raleigh, who was in fact one of Elizabethan England's first renown slavers, alongside Sir Francis Drake and Sir John Hawkins. Black slaves routinely sailed with them to the New World on their numerous privateering voyages for the Crown. Queen Elizabeth was an enthusiastic commercial partner in their burgeoning enterprise of human trafficking. Fryer notes that 'Sir John Hawkins made three trips to the America, from the West Coast of Africa between 1563 and 1567, taking with him several hundred of the Natives, which he sold as slaves' (411). Queen Elizabeth made a great fortune out of this trade and showed her gratitude by granting Hawkins a knighthood. There can be no doubt what for as the crest bestowed upon him bore 'a Negro head and bust, with arms pinioned' (411). African slavery for its part was thereafter sown into the very fabric of the English dependencies in America, with servitude as one of the cornerstones of the founding of their economic system. Fryer asserts that slavery 'became a fixture after its introduction to these Colonies, was due to prerogative of the Home Government rather than the importunities of the Colonists – especially as it was a source of revenue for the Crown' (411). Similarly, Peter Linebaugh and Marcus Rediker conclude that Elizabethan Age 'English participation in the slave trade, was essential to the rise of capitalism' (28).

Two centuries onwards, privateering ceased to become the sole means through which to justify racial subjugation as the Crown's enterprising partnerships with mercantilism expanded into new geographical territories. As the British Empire absorbed large parts of Asia and Africa into its territorial domain, it discharged millions of middle-class and working-class Britons to do the work of governing its expanding empire. From the outset, these classes were ideologically equipped with a popular, scientifically endorsed belief in their white supremacy, in contrast to that of racially inferior colonial peoples. '"Scientific" racism received its first crucial modification in the 1770s, when the British government first turned its attention to the problem of ruling a territory with "natives" in Bengal. The development of the doctrine and the development of the Empire ran side by side from then onward until the final burst of British imperialism coincided with the final burst of "scientific" racism' (Curtin 42). The rise of 'scientific' racism was extremely important for the furtherance of the British Empire and the materialisation of what we have come to know as the Western world.

The last decades of the nineteenth century constituted both 'the golden age of racism' as well as 'the golden age of the Imperial idea' (Curtin 42). Anne McClintock argues that during that exceptional period, there was 'a shift from 'scientific' racism -embodied as it was in anthropological, scientific and medical journals, travel writing and ethnographies - to what can be called *commodity racism*' a phenomenon which 'converted the imperial progress narrative into mass- produced consumer spectacles' (130). Commodity racism was crucial insofar as it came to 'produce, market and distribute evolutionary racism and imperial racism on a hitherto unimagined scale' from the 1850s onwards (130). The black body was made to exclusively hold exhibition value within this emerging system of commercial transformation. The imperial paradigm of commercial empire required that African women and men figure 'not as historical agents, but as frames for the commodity' (McClintock 141). The racist concept of an innate white supremacy, became increasingly popular throughout the 1850s, and reached the zenith of its popular approval in the period from the 1890s to the First World War. The field of human genetics was founded in that same period by Francis Galton with the express purpose of demonstrating white British racial supremacy over their colonial subjects through the use of biometrics. Galton gave rise to eugenics which was widely supported in the beginning of the twentieth century by such notable figures as Winston Churchill, Marie Stopes, Bertrand Russell, John Maynard Keynes, and William Beveridge. It also gained scientific currency in the United States through leading figures like Theodor Roosevelt, Alexander Graham Bell, and Helen Keller, and ultimately formed the scientific basis of German National Socialism's murderous racist ideology. With the discovery of DNA in the early 1950s, there came a paradigm shift in genetics from a concern for phenotype to genotype, individual to population, when it came to the scientific conception of race. What had been a concern for identifying physical traits of racial differentiation would hereto after be focused upon at the molecular level.

Scientific Freighting

Race plays an indelible role in the trajectory of western science. 'The rise of scientific racism induced not a radical shift in the characteristics ascribed to Africans or blacks in general but a reworking of those characteristics in different frames of reference' (Drescher 419-420). The black body's former ability to function as a frame during the previous era of bio-political racism has now morphed into a ladder for the ascent of a nano-political racism. Hence, the fundamental characteristics of 'blackness' have reassert themselves within the new science of cellular genomics, which filters older racial thinking through newer scientific technologies, only to arrive at a place where 'the core racial typology of the nineteenth century - white (Caucasian), black (African), yellow (Asian) and red (Native American) - still provides a dominant mould through which this new genetic knowledge of human difference is taking shape and entering medical and lay conceptions of human variation'

(Rabinow and Rose, 207). What persists throughout, is the assumption that 'races are comprised of uniform individuals' and any individual or small sample of individuals can represent the whole group' (Hawkesworth). Therein the infrastructure of genetics is built upon a foundation where centuries of global migration and miscegenation amongst these formerly "pure" types is obliterated from the collective understanding, only to be recast through an unquestioning organising principle that allows for social inequity to become once again the pre-political stuff of nature.

At a time when humanity is meant to be 'unlocking the secrets of life' through advances in genomic science, a form of biological "unknowing" is emerging that enables the 'brutality of colonization, conquest, displacement, dispossession and enslavement' to be omitted from dominant discourses around the origin and function of racial identification (Hawkesworth). Amongst African Americans in particular, DNA sequences are currently being 'used to trace individuals' "roots" to their pre-slavery origins in very specific regions of Africa, a trend which is reflective of 'the growing sense of many individuals that genetics in some way holds the key to their "identity"' (Rabinow and Rose 207). Thusly, 'race is once again re-entering the domain of biological truth, viewed now through a molecular gaze' (206).

Peter A. Chow-White argues that in the digital age information has emerged as 'the material by which we work on racial meaning' not through a departure from the paradigm of race as biology, but rather its diversion into 'a new racial formation — the informationalization of race' (1171). This differs significantly from previous forms of socially constructed racial identification insofar as, 'the "meat" has been left behind and cultural signification has become embedded in computer programs and complex algorithms, hidden from the front end, user interfaces' that correspond to a body that is increasingly being modelled as posthuman (1172). Different technologies such as CCTV, biometrics, and DNA are currently being utilised to make up the information infrastructures that have come to define racial recognition. As race is made to navigate computational routes, it becomes progressively articulated as a neutral category when filtered through algorithms, databases, data mining and the internet. 'The seemingly descriptive representations derived from information infrastructures,' quickly give way to the inscriptive as race becomes codified to virtually restructure social and political meaning within the context of globalisation and neoliberalism (1173).

In a contemporary sense, race has moved beyond the limits of a modern and liberal biological determinism, to become a dynamic mode of transmission of codes and information, its former ontological position now opened outward to accommodate the mutability of new imaging technologies, made both differential and communicative in their design. The concept of race becomes incidental when set against the superhuman radius through which a postmodern, neoliberal life is meant to be governed. In the current age, the screen has eclipsed the lens when it comes to the mediation of race as a feature of embodiment, its colonial origins largely erased from our understanding of it as a source of both truth and conflict. Racism is now approached as an after-image suspended at the end of a history we no longer care to access. In doing so we have discredited our biopolitics, only to invest and organise our bodies into what Paul Gilroy terms a 'nanopolitic' wherein 'blackness can now signify vital prestige rather than abjection in a global info-tainment telesector, where the residues of slave societies and the parochial traces of American racial conflict must yield to different imperatives deriving from the planetarization of profit and the cultivation of new markets far from the memory of bondage' ("Race ends here" 844). The digital age, in contrast, promises us access to a disembodiment of a radically different nature such that freedom is determined solely by one's power to consume, and transcendence born of the desire to consume racial and colonialist narratives from a vantage point of achievement, as opposed to knowledge and value coupled with self-expression and self-determination. Here the focus shifts from 'the *what* of race to the *how* of race', from knowing race to doing race (Chun 8).

From this standpoint, it is possible to view the racialised body as something that is able to at once 'defy the epidermal logic that has traditionally defined race' and reduce our perspective to the practicable level of invisibility thus divorcing it from the terrain of exposure (Chun 9). This is crucial insofar as our previous apprehension of race held it 'as an invaluable mapping tool, a means by which origins and boundaries are simultaneously traced and constructed and through which the visible traces of the body are tied to allegedly innate invisible characteristics' (Chun 10). Presently the situation of race is entirely reversed and racial consciousness is submerged within the body for which the appearance of race must become indexical. Within the schema configurations of race become cultural artefacts, and race itself a placeholder signifying our evolution from a past defined by limitation, to a future of boundless enterprise. Racial and ethnic distinctions therein can be reduced to disparate social locations and distinctive histories, rather than any institutional form of enforced segregation.

Race no longer has to offer itself as certainty, but rather as a form of property, which can then be grouped and regulated by newer forms of disciplinary power that correspond with 'new kinds of visual culture which re-incarnate racial differences and make them meaningful in unprecedented ways on a novel global scale' (Gilroy "Between Camps" xiii). In a post 9/11 world where (neo)imperial domination once again has become the normative politic, it is perhaps more important than ever to look at the tradition of dissident and critical thinking offered by those ethnic groups subject most painfully to its previous incarnations, nor should they forget 'their historic responsibility to act in solidarity with the post-colonial movements for justice and human rights that are flowing out of the global "south" and composing a new planetary network in pursuit of a more thoroughgoing democracy than was offered earlier in colour-coded forms' (xiii). This is something extremely difficult to accomplish given that so much of recent black cultural production 'has been commodified and exported to the rest of the world, namely as an effect and vector of a new empire that remains profoundly attached to racialized notions of power, order, justice and entitlement' and furthermore, that it does so at a time when black communities are increasingly divided along class lines that splinter their once unified economic and political interests and yet remain beholden to their poorer members to supply their sense of what constitutes their racial authenticity (xiv).

Nationalising Dispossession

These concerns date back to the nineteenth century where it can be argued that at the heart of this black politics were efforts to assert the rights to citizenship to combat the horrors of coercive labour and enslavement. In the twenty-first century, black nativism takes on an additional meaning within a highly globalised society, where notions of affective belonging and material citizenship resonate deeply with constructions of past historical trauma. In the context of the United States, 'there are blacks who say that [if] you don't carry the psychological burden of slavery, or growing up in Harlem, or the south side of Chicago as descendants of slaves' you fail to qualify as an African American (Barrio-Vilar). This attitude not only denies the hybrid and multicultural experience of African Americans in particular, and Americans in general, but equally the complex racial identities of African diasporic peoples who have arrived more recently, are of mixed race background, or haven't spent their childhoods primarily amongst African Americans. It is not just colour but class that determines one's belonging, as those with privileged backgrounds are often denied full entry, as well as those whose racial background was not founded in American slavery. It is not sufficient to be African or Afro-Caribbean, rather one's identity must be underscored by its connection to the history of the American society writ large, and thus intra-racial tensions are felt at the level of race, as well as migration and citizenship, especially in light of the constant flux of disaporic communities from Africa and the Caribbean which must find their place toward a broad definition of racial phenotype. Black nativism has always been a by-product of dispossession.

In the United States blacks were never considered equally worthy of citizenship as compared to whites. It was this dilemma of legitimacy that powerfully shaped the language and tactics of black politics, which developed a rhetorical language of black nativism to further the assertion that birth on American soil and the contribution of one's ancestors to the American nation, had won African Americans the right to be citizens of the United States. Many African Americans believed that the elevation of their race was dependent on adopting economic and commercial schemes similar to those of whites. In the nineteenth century, this extended to an ambition for African Americans to colonise Africa, and to exploit it economically in the exact manner in which white Americans and Europeans once did, in order to generate an African American counter-imperialism (Adeleke 55). Those in favour of this approach alleged that the hardships of slavery had been in some sense a 'progressive experience' for blacks, insofar as it 'bred' enlightened African Americans, who when put to the test of their mental and physical endurance, had managed nonetheless to develop and to increase their numbers, such that is was now their destiny to lead the societal regeneration and economic cultivation of Africa in a way that was superior to whites, securing it as an economic and political power base for blacks fully outside of the domain of the Americans and the British (56).

The lack of integration of blacks into the wider American community, and their poor prospects for economic advancement within the context of American culture, meant that their African consciousness would prevail for centuries to come as a dual identity and as a buffer to white American prejudice. At the same time, Tunde Adeleke argues, their self- conception as Americans was 'infused with values that affirmed white supremacy' (116) Moreover, the ideology of 'racial hierarchy' that had underpinned slavery and segregation would denigrate the image of black selfhood and black cultural heritage long after these institutions had been formally dissolved. The European materialist and technologist view of what constitutes civilisation required African Americans to participate in the dismantling and subverting of African sovereignties in order to achieve class ascendency through what essentially amounted to black imperialism. In doing so they wilfully ignored 'the global character of racism', in order to be afforded 'recognition as partners in the spread of civilization' (134). It was that harsh bargain that created lasting and painful ambiguities within the relationship held between American blacks and their African counterparts that exists to this day. That is the context in which we become obliged to appreciate the deployment of African American culture as a diplomatic and military instrument—something that captures and can amplify the restless, intoxicating appeal of "team America" as well as the fantasy that all the distinctive freedoms and voracious patterns of consumption enjoyed by the most privileged people in the United States could be enjoyed by everyone else on earth (Gilroy, Race and Racism in "The Age of Obama," 10).

Digital Natives

Race within this scenario assumes a new ontological function that transcends nationalism and joins with a new politics of information capable of powering ever greater forms of control of who can be recognised and what can be known. Who is native to these virtualised, postcolonial settings, relates once again back to 'the historic pressures of racial hierarchy and ethnic absolutism, make judging who might be worthy of recognition as fully and authentically human' and who gets relegated to the vagaries of alterity a matter of precarious deliberation (Gilroy, Race and Racism in "The Age of Obama,"8). These processes still rely upon the institutionalised double standards previously characterised as black and white, but add to them a nuance that caters for a cultural supremacy to tip towards the valorisation of privatisation, financialisation and militarisation. New patterns of identification have established conflict by other means, to be waged on new maps prepared by the revolution in digital communications.

The term 'digital natives', i.e. the current generation of young people as the first to have always had access to the Internet, incorporates within itself a level of economic privilege that is far from universal in terms of its applications and its divisions across racial and ethnic lines. The Internet in the course of a single generation has been able to achieve its own unique form of sovereignty and territorial command. The digital seeks to capture control in a way that makes our becoming matter along continued lines of problematic social inequalities and racialised lines of belonging that persist in to this, the West's latest narrative of manifest destiny. As the digital encroaches upon our various social institutions, we as a people become increasingly subject to meditatisation, our bodies routinely penetrated by new platforms, devices, and applications that command our constant attention and assemble us into networks that compromise the very bounds of reality. Pervaded upon in this manner, the digital has become surrogate to our thinking and structure to our knowing, reformatting the ways in which our bodies assimilate productivity and introduce consumption. Information has become its own methodology and through it we must find ways of operating as bodies to reflect current media realities. In his seminal book, *Between the World and Me*, national correspondent for *The Atlantic* Ta-Nehisi Coates reiterates to his son, a disturbing encounter he had with the host of a popular news programme:

> The host was broadcasting from Washington D.C., and I was seated in a remote studio on the far west side of Manhattan. A satellite closed the miles between us, but no machinery could close the gap between her world and the world from which I had been summoned to speak. When the host asked me about my body, her face faded from the screen, and was replaced by a scroll of words, written by me that week.
>
> The host read these words for the audience, and when she finished she turned to the subject of my body, although she did not mention it specifically. But by now I am accustomed to intelligent people asking about the condition of my body without realizing the nature of their request. Specifically, the host wished to know why I felt that white America's progress, or rather the progress of those Americans was built on looting and violence. Hearing this, I felt an old and indistinct sadness well up in me. The answer to this question is the record of the believers themselves. The answer is American history (Coates 5-6).

Coates' digital encounter is superlative in a number of respects. Firstly, in his acknowledgement that racial differentiation itself implies a remoteness of one body from another, a fundamental divide that requires interpolation of one kind or another to correlate it. It is the host, which must extend such a gesture of power to call up Coates as the very embodiment of her query. Within the space of this request, she, on the other hand, literally gets to disappear. Her body faded seamlessly into his recent textual expression. The host reasserts herself at this moment through her voice, which carefully mediates Coates' language for the audience, her voice, not his, acting as command. This pattern of usurped authority continues, as the host now focuses her attention from Coates' words, back to his body, which exists here in a conditional sense, insofar as she is able to intelligently and likely without realising, instruct him to respond to a general condition of 'blackness' and group it with other bodies that have been 'mocked, terrorized, and insecure[d]' (Coates 55). Perhaps '[Coates'] being named "black" had nothing to do with any of this; perhaps being named "black" was just somebody's name for being at the bottom, a human turned into an object, object turned into a pariah' (Coates 55). Coates, as a perceived social outlier, is asked to comment on his 'feelings', regarding the progress of white America. The feeling he is meant to draw upon is a deep well of 'physical pain and exhaust[ion]' attendant to his representation within America, who made his grossly objectified body transmit the miserable burden of association with 'looting and violence' spanning some four centuries (Coates 55). The host herself making that record over once again inside this particular broadcast, echoing the racialised beliefs that have come to bear not just on America's history, but its present re-presentation of itself as a contemporary 'postracial' society, where nonetheless race has emerged as an informational commodity, and as an award offering more recreation to its audiences, than to its subjects. Race as a technology has been routinely tasked with generating specific forms of oppression and discrimination that resonate with the context of their application.

Within the contours of contemporary neoliberal ideology Coates is made to comment on 'blackness' as an affect of personal identity, rather than something that infringes the domain of collective American responsibility. In some sense race has been held back by digitisation from its progress, as a distinctive narrative, that can be imagined, mapped, and visualised within the wider index of American identity as something more than 'a photonegative of that of the people who believe they are white' (Coates 42). Coates' carefully staged interaction with his [white] host powerfully demonstrates how 'digital interactions provide key sites for the delineation of hierarchies of belonging, and the expansive rehearsal and contestation of racializing discourses, tropes and rationalities' (Titley 42). In effect Coates' body, made proximate to his host only via a satellite link is nonetheless signalled to expound upon a racialised form of history that carries within it all prior incarnations of misappropriated black embodiment, including those in other popular media formats.

Taping Suspects

In the second half of the 1980s, America began to enter a new era of racial transcendence. This shift coincided with a sudden deregulation of financial markets, but perhaps more significantly the advent of electronic, screen-based trading. At this time race as a commodity transformed radically from a 'modern object' to a 'postmodern subject' as the financial market emerged as the arbiter of all political possibilities (Iton 125). In his highly nuanced historical survey of black Atlantic politics, *In Search of the Black Fantastic: Politics and Popular Culture in the Post-Civil Rights Era,* Richard Iton argues that by the late 1980s, the transgressive scope of the black counterpublic of the civil rights era had reduced down to a radically disaggregated screen on which to project an emergent black 'superpublic'. Race within the new era could be released from its political modality, and crucially resituated towards an appropriative function to be subsequently channelled through a vastly expanding number of electronic platforms being devised at this time. Political descent relinquished crucial ground to an aesthetic appropriation of 'blackness' with 'the advent of videos and music video channels, [and] videocassettes' and 'video games' in the 1980s and later, digital videodiscs (DVDs), dualdiscs (combining CD and DVD content) and the internet' in the 1990s (Iton 104). When brought together these multimedia formats, 'simply overflowed the previously existing networks of intercommunity discourse' (104). Through these new media technologies, protest and resistance were co-opted into commodities that could then be resold back to black communities now understood as a growing segment of the economy. Black communities now found they had to confront a vastly expanded number of platforms through which to view depictions of contemporary black culture. At the same time, white communities could now critically judge and extract pleasure from an array of newly highlighted and hyper visible depictions of race and ethnicity.

The preponderance of media narratives documenting black success that emerged in the late 1980s contributed to the perception that

America in particular had entered a new era of racial transcendence. Radical inequalities between whites and blacks when it came to wealth, housing and education could now be obscured through a determined focus on individual effort, perseverance and achievement. Within this portrayal, it was assumed that those who could not escape homelessness, destitute urban neighbourhoods, poverty, public housing, and welfare dependency had surrendered to the cultural deficiencies that surrounded them, their dismal situation having nothing whatever to do with political, or economic policies. Rather it was a matter of taking personal responsibility for rising above one's circumstances however blighted and make one's way toward phenomenal success against all odds. The fact that black achievers, have 'attracted a majority white audience is interpreted as a major advance in American race relations' (Peck 100-101). Perhaps white's embrace of black celebrities reveals more about the power of television than the actual state of American race relations, because the United States remains highly racially segregated. Many white's primary encounter with African Americans comes through television - a medium that presents a dangerously schizophrenic image of black America split between 'super-successful and largely admirable (not-all-that) black' middle class and an underclass of 'dangerous (all-too) black perpetrators' (Peck 101).

Television allowed blacks to be received into the homes of white Americans on a regular basis, without posing the challenge of actual an interracial encounter. This encounter nonetheless, 'give[s] white Americans the sensation of having meaningful, repeated contact with blacks without actually having it' - phenomenon Leonard Steinhorn and Barbara Diggs-Brown term 'virtual integration' (146). In their view, 'virtual integration enables whites to live in a world with blacks without having to in fact do so. It provides a safe intimacy without any of the risks. It offers a clean and easy way for whites to establish and nourish what they see as their bona fide commitment to fairness, tolerance and colorblindness' (146). They argue, 'it is through virtual integration that 'whites have made room in their lives for black celebrities' and take their affection for such figures 'as evidence of their own openmindedness and as proof that the nation isn't so hard on blacks after all' (146). By contrast, Steinhorn and Diggs-Brown point out, this same colorblindness is 'almost unattainable for blacks in the real world' (146). Colour in this way can be virtually diminished provided that you are willing to see integration as a product of private projection, rather than public imposition.

Such a stance toward colour-blindness does little to challenge the status quo of power relations between white and black. Through the advent of neoliberal economics, a system predicated on institutionalised inequality for all, people of colour are now ostensibly admitted to the field of competition to savour the spoils of free market capitalism. Similarly, a new era of apparent 'colourblindless' when it comes to the recognition of wealth, allows inequality to assume the guise of competition and diversity the guise of meritocracy, and in so doing promote a virtuality of social justice that never materially integrates with the reality of institutional discrimination. Within such a construct, Walter Benn Michaels observes, 'it's basically OK if economic differences widen as long as the increasingly successful elites come to look like the increasingly unsuccessful non-elites. So, the model of social justice is not that the rich don't make as much and the poor make more, the model of social justice is that the rich make whatever they make, but an appropriate percentage of them are minorities or women' ("Let Them Eat Diversity").

As we entered the 'Age of Obama', neoliberal black attainment was made synonymous with its white counterpart, insofar as it was understood to derive from having extraordinary talent, exerting a strong will and possessing an unrelenting work ethic. This racial equivalency in terms of recognising success required the strategic disavowal of an 'endemic' blackness, in favour of an implicit affirmation of cosmopolitan whiteness, which now functioned as the new contour of social and economic power. As race was permitted to become something non-essential it became possible to transfer its value from a public to private concern, complementing a new consumer orientation to bodies of colour. Racial difference would no longer be derived from a category, but rather projected as an event precipitated through the actions of an individual.

Appropriating the Products of Empire

Racial perception is something that we hear as well as see. On this basis, it is possible to conclude that the governance structure of mass media is multisensory in its criteria of subtle exclusions. This scenario leaves black performers open to exploitation through the allocation of extra work to them to achieve comparable outcomes to their white on-screen counterparts. In contrast to them, they have to play to achieve both character and race. It is implicitly understood that they must refrain from criticism of their employers and indeed the greater institutions of Hollywood and finance capitalism that shamelessly promotes them as outsider superstars i.e., individuals who with seemingly few resources to draw upon still 'made it' in the industry. This approach essentially ignores the existence of systemic racism, and presents any discussion of racialised typecasting as anachronistic. Regrettably the ostentatious presentation of the rise of the black British actor in Hollywood guarantees little progress at the level of institutional norms informed by a long history of racial discrimination. Although it cannot mean the same to people who have migrated from Britain, Nigeria, or Ghana it still very much bears some tangible, past relationship to the racial hierarchy of the former British Empire. The persistence of that hierarchy in contemporary Britain was indeed the very thing that eventually brought them to Hollywood.

Black Britain's recent arrival in the United States is one that was digitally mediated, and yet on another level it relied on a much older convention of privileging an international form of blackness over its domestic counterpart, making the acting careers of British actors of Afro-Caribbean background in Hollywood a more viable potential as compared with African Americans. The slavery era colour lines that continue to define and restrict African American on-screen talent, do not apply to their colonial diasporic fellows whose contemporary American cultural capital derives from an entirely different, cultural, political and economic paradigm of racial encoding. Racial privilege is routinely attached to location and international blackness afforded through the British accent grants specific racial licence for international blacks to be perceived as proprietarily superior to domestically cultivated blacks. In a neoliberal economy, this phenomenon is credited to their having superior theatrical training, rather than a hierarchy of racial identity that has persisted from a much older Atlantic economy.

Black Britons ascribe their exile to Hollywood, to institutional forms of racism that remain as the toxic residue of Britain's imperial and colonial past. Their remedy is said to reside with a triumphant capitalism that will allow these actors to grow rich abroad based on a combination of hard work and the recognition of their innate ability. There is something glaringly omitted from this equation, insofar as these actors must professionally adopt false personae in order to achieve this feat through the faultless mastering of an American accent. There is something telling in this insofar as this implies their adoption of a categorical designation is associated with class, perhaps more so than race. The British accent in America is associated with the upper classes, therefore an American audience might be confused by a black actor speaking with a British accent, because they have on one hand difficulty conceptualising the ethnic diversity of the UK and on the other adopting a belief that blackness can be something synonymous with a superior lineage of any kind.

Furthermore, Americans remain almost wholly ignorant of the history of Britain's involvement in the transatlantic slave trade that made blacks and other ethnic minorities part of Britain's national history dating back to the same historical period of the eighteenth and nineteenth century, which incorporated them into societies in the United States and the Caribbean. Evidence of that history is seldom depicted in the productions Americans are seeing coming over from a contemporary Britain, which continues to promote the perception that black people simply did not exist in the UK until after World War II; a concept that conveniently absolves the culture of responsibility for the violent cultural uprooting and racial suppression that was concomitant to a British colonial rule now entering its fatal period of decline. American ignorance of these phenomenon, allow black actors with British accents to negate rather than activate racial stereotypes in their auditions, and later to integrate that skill into their working identities by portraying black characters that look black, but otherwise conform in appearance to white value systems therefore making them racially palatable to audiences. These black British actors remain aware that they are at all times giving a racial performance. Most are not cast in roles where they are called upon to speak in a 'racialized' dialect. Rather majority of their roles are situated amongst the middle class so that he's or she's is not exclusively and explicitly encoded as conventionally 'black'.

Ironically, the work that is generated for them once they arrival more often than not contributes 'to the fantasy version of African-American culture that's been exported to the rest of the world' (Yancy and Gilory). That portrayal of 'blackness derives in large measure, not from the ordinary lives of black folk', but 'from the dream worlds of global consumer capitalism (Yancy and Gilory). This imaginary form of 'blackness' is one that is 'heavily mediated by the Internet and social media, and its dismal effects are compounded by the general crisis of political imagination' (Yancy and Gilory). In the twenty-first century blackness remains to some degree a marker of visible difference from an implicit white norm. Its recent elevation to desirability as something that may be highly prized, performs a particular function insofar as it acts as 'a sign of timeliness, vitality, inclusivity, and global reach' of Western capitalism (Gilroy "Between Camps" 22). The meaning and status of racial categorisation has been unmoored by the recent phenomenon of 'placeless development and commercial planetarization' as the world itself flattens under the pressure of the global market, so too do substantial linguistic and cultural differences, allowing race to become the agent of continuity and overlap, as opposed to functioning as a major means of differentiation (23-24).

Policing the Representational Crisis

The anxiety around the admissibility of black humanity into visual representation in no way diminished with the advent of a postmodern black popular culture in the late 1980s and 1990s. Rather its sudden appearance and stylistic particularity was construed as a violent takeover of a predominantly white popular imagination in an unprecedented way, insofar as this version of blackness overtly celebrated the market process and willingly proffered its historical struggle up to the forces of commodification. Racial hierarchy and marginalisation were not significantly compromised through this commercial development. Rather such hegemonic ambitions merged with 'a range of gatekeeping responses from those committed to restricting the circulation of certain kinds of information within black communities and maintaining "order" and exhibiting respectability' (Iton 104). The black community was thus charged with policing itself. The implication being, that if it failed to do so punishment would come in the form of hyper-violent white pseudo-military civic policing.

This threat had been issued against African American communities from the mid-1960s onwards, when American police began to employ tactics and technologies for quelling 'racial rebellion' at home that were directly adapted from 'military manuals from the United States Agency for International Development in Asia, Africa, and Latin America' (Roy, Schrader, & Crane 140). Police were trained to adapt methods of pacification and counterinsurgency, developed overseas during imperial warfare operations, to subsequently formulate derivative urban policing strategies deployed throughout the United States. During the Vietnam war, game theory, a corollary of cybernetics, became a favoured means of predicting and managing chaotic scenarios, which not only reconfigured conventional military tactics, but also acted as a precursor to 'technologies of scenario modelling, surveillance and data analysis' that have direct bearing on 'civil and economic spheres of governance' rapidly evolving from that era to include monitoring the conduct and participation of differential subjectivities (Vishmidt 261). Within such a scenario it is possible for social control to become compatible with 'both egalitarianism and particularity ("diversity")' and for that situation to exist within a greater system of 'overall enclosure' that gives premise to 'the constitution of social life by capital forces' (262).

This condition displaced modern forms of racism, that relied upon ascriptive biases characteristic of Jim Crow and pre-1965 black life and resituated them to conform with a status based on a certain kind of practicable achievement, the ability to seize and exploit the evasive potential of the performative. The predominance of the visual in popular culture and elsewhere gave licence to an emerging fallacy 'that black bodies might escape sanction if they are committed to the presentation of non-blackness- or, more accurately, acceptable and assimilable form[s] of coloredness' (Iton 104). This misapprehension would compromise the authority of a previous black politics, which raised alarm in the wake of a field of representation that portrayed them 'as being inherently unruly, surreal, outside the frame, and in need of discipline' (Iton 105). It was the same characterisations that attracted the attention of the new visual apparatus that now sought to engage, in particular, with black subjects, and to lock on to black visibility and blackness as a unique driver of commercial productivity.

The emergence of a black 'superpublic' in the postsocialist world of the 1990s allowed for a reassertion of the colonial gaze, which had always been present as a default setting in white America, to become awesomely restored through the lens of an American post-Cold War global hegemony. The collapse of the Soviet Union and the abolition of apartheid in South Africa marked the end of the resistance to America as the world's sole remaining hegemon, allowing the discourses of the postcolonial, postpolitical and postracial to emerge to comprise the articulation of lingering institutional antagonisms. Iton discerns, 'in the absence of a robust, transnational, and self-defining black counterpublic (in the form of allegiance to a black nationalism or resistance to apartheid) or more accurately, the overlapping cluster of formations that constitute "black politics" - black performances might increasingly bend and be bent to serve, and be read as seeking to engage, seemingly new but frankly familiar masters in the context of this different conjuncture' (Iton 128). This new era of racial performativity rapidly adhered with an 'internalization of the expectation that one is always potentially being watched' (Iton 104). Within such a heightened state of readiness it became imperative for the racialised subject to perpetually 'perform even when one is not sure of one's audience (or whether in fact there is an audience)' and to therefore, 'give one's best performance and encourage others to do the same' (Iton 104). Within this precarious speculative economy, individual performance had to coincide at every point with prevailing notions of authenticity and sincerity, because it was assumed it might be later judged according to the equivocal standards of televisual mediation.

These actions would be made visible through the excessive, or spectacular appearance of their bodies through the medium of various new media technologies, which in effect 'raced' the politician, footballer, television personality, or ordinary citizen as a consequence of their behaviour. Thereafter, 'appearing' black would be a punishment for having in some way breached the signifying parameters of inclusion within a colour-blind society. This neoliberal ethos emerged in the 1990s when African American bodies began to appear on videotape as a medium through which to exorcise national trauma. Some of the most notorious videos of that era include 'former Washington D.C. Mayor Marion Barry video-taped crack smoking and subsequent arrest; the Clarence Thomas hearings; Mike Tyson's rape trial; Magic Johnson and Arthur Ashe's televised press conferences about their HIV and AIDS status and, of course, the Rodney King beating' (Alexander 79). If whites were the primary consumers of these mediated spectacles, for blacks it made a disturbing 'product' of their lived reality of violence and surveillance, reifying the assumption of their position as that of 'countercitizens,' gravely incompatible with the prevailing [white] perceptual order (81).

Such scrutiny intensified with the proliferation of handheld video cameras and surveillance cameras that transferred formerly military preoccupations into the conscience of a civilian population, thus generating a heightened valance of self-awareness and interiorised paranoia amongst the general population and in particular, already targeted minority communities. This has certainly been the case with mobile device recordings and social media instant distribution of such footage over the past few years. These recordings seek to challenge long held beliefs around race, privilege and power by acting as digital sites of resistance against racially motivated violence and toward any acceptance of the notion that America has achieved a post-racial society. White police officers, targeted minorities and bystanders all have a stake in controlling the new narratives around race that have subsequently arisen with the proliferation of surveillance video now increasingly captured on civilian smartphones, dashboard cameras and police body cameras.

These mobile technologies have become the latest modes through which to convey the nation's collective public reckoning. The documenting of the violent loss of black life online, and its subsequent existence in virtual perpetuity, within the context of a 'sharing economy, allows black death to become something to be consumed in 'overwhelmingly white online spaces' such as Facebook, Twitter, and

Google (Sutherland 38). As a social, cultural and political commodity black death becomes its own property within the territory of an ongoing white supremacy, which gets to ultimately master these images. Such images therefore seldom become the basis for social justice, but rather occupy a different position altogether in digital space through there infinite potential for appropriation in an ideographic system of structuralised white supremacy. What is left to picture here is what Judith Butler refers to as 'the racism that has become ratified as public perception' ("What's Wrong With 'All Lives Matter'?").

Such ratification relies on the advent of what Raphael Foshay refers to as the 'dubject,'or digital subject doubled and spaced, 'translated' as it were 'from the site of its individual body to the mediated space of representation' (129). Such a self, is 'committed to its own recording', and as such becomes an entity 'self-dubbed and doubled' into the form of 'a doppelganger self whose "live", corporeal presence becomes radically supplemented' through 'its different and distributed embodiment in recordings and representations' (129). The dubject exists within both a postmodern and postcolonial context, and yet remains somewhat atavistic, insofar as he is often reduced to a white, male identity formation who is plotted to achieve transcendence as a disembodied, posthuman mind within a globalised milieu that otherwise routinely dictates physical violence directed against his class, race and gender subordinates. His digital privilege is predicated on the intensification of their corporeality.

The internet itself operates very much as an extra-state entity, insofar as it insists on identification and verification before occupation is granted to the dubject. Those entering must concede to aligning themselves with the requirements of neoliberal capital and the surveillance state to constantly report on themselves and to confirm their dual whereabouts. The territoralisation of the digital has a great deal in common with earlier historical periods of active colonisation. What 'we are witnessing [is] the new colonial expansion – from reality to virtuality' wherein digitalisation is but another version of the coloniality of power and being that has been at work for centuries (Foshay 36). This latest version of it is about finding an outlet for colonality to redefine itself through a new landscape of violence, where the enemy is mostly unseen, there are no obvious goals, no distinctive frontiers and societies through which to chart these encounters. This version is not even ostensibly about bringing progress, but rather about exerting various levels of meaning and control on the unknown and unknowable.

Encoding Race

The informationalisation of race is at work within a diversity of settings including law enforcement, biomedical research, insurance and marketing, and manifests its power through various prosaic modes such as video surveillance, DNA collection, online patient databases, and self-expression on social media. The latter have emerged as specific modalities of race within themselves, where 'casual racial banter, race-hate comments, "griefing", images, videos and anti-racist sentiment bewilderingly intermingle, mash-up and virally circulate' (Sharma 47). Social media has allowed racial encoding to relocate itself from the public sphere to private and personal spheres where identity-based biases can resituate within the context of a hyper-mediated culture. This allows race and racism to be narrated as the product of a series of individualised choices, rather than the consequence of collective, institutionalised practices. 'Subsumed under a language of choice and detached from power and structural disadvantage, the denial of inequalities can be seen as part of the broader cultural retrenchment, nostalgic vitriol and coded xenophobia sweeping the political terrain' (Redclift 580). Very old forms of race and racism dating back to European colonisation are migrating into the digital space, at the same time as new digital formations are being developed that correspond with contemporary neoliberal globalisation. Such digital relations come into being in and amongst the transition between past cartographies of movements and conquest and present forms of uprooting and migration. As a result, the digital production and circulation of race is not just a simple extension of offline racialising discourses and practices, but is generative in its own right, as a means and method for race to journey and amass across new networks of understanding.

Race today occupies a contested territory, held between the former epidermal and future digital that acts as a site where certain bodies are permitted and others made out of place. That critical intersection is where we locate our contemporary consciousness with regard to the possibilities for movement and contact, between those who get to freely choose their claim to their ancestry based on their not looking the part and those who had no choice in that discovery, nor the persistent visibility of their encoding. Within the shadow of this economy lies the concept of branding which has its own darker history for African and Caribbean peoples. Hortense Spillers identifies branding as one 'of the key technologies of the trans-Atlantic slave trade,' (Browne, "Digital Epidermalization", 138). The violent appropriation of the black body as a seized territory of cultural and commercial manoeuvre is a centuries old practice. Enslaved persons were 'physically encoded with numbers and letters that identified them as being part of a particular ship's cargo. This practice also worked as a system of identification that enabled surveillance' (139). The brand, applied directly to the skin of the individual was 'often the crest of the sovereign, was a stamp of commodity, a signifier of bondage' and of the geographically distinguished 'relation of the body to its said owner' (139). This encoding worked at both the level of economy and governance to ensure that certain bodies were permanently registered inside a greater system of mass state and corporate tracking of populations.

Simone Browne's ground-breaking work on this subject concludes that while current biometric technologies and slave branding are not one and the same, 'in our contemporary moment where "suspect" citizens, trusted travellers, prisoners and others are having their bodies informationalized by way of biometric surveillance' it cannot be denied that 'the historically present workings of branding and racializing surveillance, particularly in regard to biometrics' speak to a relationship between branding and the body as a system of identity that persists through its present virtual encoding' (251). Within this system, 'non-white' bodies are disproportionately inventoried by way of a biometric surveillance that monitors, tracks and classifies races for specific figurative inclusion within an increasingly techno-ontological order. Such bodies were and remain predetermined as candidates for control and censure, their heightened legibility functioning as both a means of knowing the body and specifically containing it. Such containment could not be achieved but for the fact that individuals over a course of two centuries wittingly, or unwittingly provided categorical information about themselves to state and private interests capable of owning and storing that information in vast official archives and in a postwar context within large-scale databases.

Denoting Territory

The infrastructure for this, what Donna Haraway famously referred to as a new 'informatics of domination', was initially devised during the Cold War in Silicon Valley (28). One of America's desired outcomes of the Cold War was to re-engineer the social and political minds of individuals to be receptive to 'new notions of a universal humanism' (McElroy 2). The new mind set would be predicated on an acceptance of science and technology as the arbiter of essential gateways through which individuals could maximise personal freedoms. Militarily employed scientists and psychologists would become the societal experts charged with culturally legitimating this mind set beginning in the late 1940s. The freedoms promoted from its inception were differentially coded according to somatic and geographic boundaries which carefully separated the aspirations of white suburbanites from their ethnic urban counterparts. Though separate, each imaginary held within it a principle desire for security and containment as presumably 'democratic' features of spatial management.

Unlike San Francisco, Silicon Valley's suburban regions, were expressly designed from their inception to conform to corporate and military necessity. In terms of a controlled spatial logic, its pioneering 'high-tech firms' were consciously 'surrounded by prefabricated US Army-erected houses' (McElroy3). Their proximity to one another would foster the ideal conditions for 'the emergence of what Nick Dyer-Witheford describes as 'futuristic accumulation', or 'the commodification of publicly created scientific knowledge, which via copyright and patent, becomes privatized as intellectual property for the extraction of monopolistic technological rents' (Dyer qtd. in McElroy 3). Excluded from these new divisions where the heavily racialised geographies upon which these regions were initially founded as sites of abundant agricultural extraction.

The newly valuable commodity of silicon redefined not just the industrial character of this area, but also its racial, class and gender hierarchies. The once lauded manual labourer lost favour to a cleaner, lighter type of worker whose fortunes where no longer allied with the land, but rather to technology. High-tech companies were determined to import in their place, a 'better class of worker,' rewarding them with higher salaries so that they would be in a financial position to inhabit the prototype [white] suburban housing concurrently being developed throughout the region (Findlay 132). In 1968, the San Jose News, proudly proclaimed that 'Santa Clara is white collar… and 93.6% of the county population [is]white' (Choate qtd. in Park and Pellow). This was the dawn of the post-Fordist era, where manual labour was rapidly degraded and a new era of office based administration was promoted as the future of gainful employment. The factory that had been so crucial to the fabric of the early twentieth century American expansion, had now taken on a corrupted appearance in the latter part of the century, as the site of left wing social and political agitation racially coded as 'un-American' and contaminated by ideas brought in presumably from 'elsewhere'. Social and political movements 'such as the Black Panther Party, the Brown Berets, La Raza Unida, and the San Francisco State strike of 1968' struck fear into the hearts of white suburban dwellers south of the cities San Francisco and Oakland, and in more affluent neighbourhoods of the city where blacks were kept away through practices of redlining (Park and Pellow 415).

In December of 1961 the Ford Foundation initiated its first 'Gray Areas' programme. The programme was ostensibly 'aimed to disrupt social unrest', thusly 'the program discursively abdicated "race" with "poverty" to mitigate "juvenile delinquency", gang violence, and "human problems in the city" through assimilatory projects of "human engineering"' (McElroy 5). Oakland as a city was to become a 'laboratory to study various methods of social intervention', which 'combined cultural and educational programming, social science and public health research, and increased policing and management of delinquent and "pre-delinquent" youth' (Roy, Schrader, & Crane 142). Erin McElroy asserts that 'what such grammar elided was that social unrest was quelling in

Oakland not because of juvenile delinquency, but because white homeowners grew racialized anxieties about the rise of Black homeownership' (5). At the time, the term 'Gray Areas' was a euphemism for neighbourhoods becoming predominantly African American. Oakland became the first of six urban sites where the program was to be implemented. Though the population of Oakland as a whole declined between 1950 and 1960, the black population nearly doubled. As white families left Oakland and built its prosperous suburbs, black families began to buy and rent homes beyond the boundaries of the West Oakland ghetto. Fears of unruly, criminal youth were closely linked to the perception that these newly arrived migrants threatened the stability of the neighbourhood. Within that perception stood a belief that the 'ghetto and colony ...are inextricably linked in a global formation of power' (Roy & Schrader 140). The 'domestic pacification, the program proffered, was a means of what Robert McNamara would later be described to the World Bank as "defensive modernization." These Cold War technologies were first tested in the Valley's own racialized backyard before being employed elsewhere' (Ayres qtd in McElroy 5).

Race became in these instances a Cold War technology in its own right that would bear implication that travelled on similar circuits to the United States geopolitical strategies, where the continent of Africa and the Americas were seen as ideologically hostile environments to the freedoms of American entrepreneurialism, Asia was viewed as a territory far more receptive to techno-capitalist assimilation and thus Silicon Valley's racial codes mutated to incorporate Indian and Chinese engineers into their burgeoning California ideology. McElroy maintains that 'such contexts of Cold War subjectivity helped fuel myths of Silicon Valley culture as one of meritocracy, multiculturalism, and postraciality towards the end of the Cold War and into the postsocialist era' where what mattered was your mind and not the appearance of your body (McElroy 5). This desire to exploit brainpower was in itself a form of assimilation, insofar as some Asian American enclaves in the Valley were racialised differently than Others, and therefore became its own testing ground for how racism would eventually play out in a post-Cold War era neoliberal Silicon Valley. It remains the case that Asian American males are overrepresented as low wage engineers in the technology sector, and Asian and Latino female migrants are the ones who are tasked with the toxic manual assembly of the semiconductor chips that go inside Silicon Valley's physical commercial products. Within this economy people of colour are compelled to differentially organise and hybridise themselves in concert with technology, as a means of giving expression to the recombinant processes that translate racism and colonialism into a postwar racialized architecture. Such an architecture was built to be conversant with a cybernetic world engineered to perpetuate the myth of its own universalism.

Liberal Feedback Loops

Poe Johnson makes the case for race as an enduring 'technology' that has its beginnings in America as far back as the mid seventeenth century, 'when Virginia property owners created laws that separated out the Africans from the Irish, then simultaneously objectified the tool they had just created' (368-369). Race as a technology becomes synonymous with the founding of America, through a multilateral system of trade linking Britain to West Africa, the West Indies, and the colonial United States. In a contemporary sense, Johnson argues that the United States acted and still acts as a founding laboratory for constructing racial technologies that are primarily animated through the progress of media technologies. The mediated transmission of differential cultural narratives had played a disproportionate role in reifying race and racism. Johnson argues that in our present age this takes place through what he calls a 'cyborgnetic media,' a portmanteau of 'cybernetic' and 'cyborg', he uses to illustrate 'that racialized bodies are biological technologies whose positionalities interface with the participatory modalities of contemporary digital culture' (Johnson 369). 'Whiteness is technologized as free and capable of agency and autonomy. Blackness on the other hand is technologized as technology: a tool for utility. To be black is not to be a person, it is to be both a servant and exist in service. Moreover, any resistance to this technological position necessities that the machine physically, visually or semiotically self-regulates' (Johnson 369). This self-regulation would be something that would be highly prized amongst a group of scientists and psychologists poised to develop possibilities of how the democratic mind and its liberal principals could be globally disseminated, through an unprecedented merger of technology, capitalism, and politics to draft a new script for universal humanism proper to the twentieth century and beyond.

The MIT mathematician Norbert Wiener, the renowned theorist of human-machine interaction, would be at the forefront of this endeavour to construct social spaces in which individuals could exercise maximal freedom. Implicitly this required the containment of others who would function as the material tool and rhetorical technology for that societal transcendence to be realised. 'If owning oneself was a constitutive premise for liberal humanism, the cyborg complicated that premise by its figuring of a rational subject who is always already constituted by the forces of capitalist markets' (Hayles 86-87). Katherine Hayles, concedes that Wiener's writing indulges 'in many practices that have given liberalism a bad name among cultural critics: the tendency to use the plural to give voice to a privileged few while presuming to speak for everyone; the masking of deep structural inequalities by enfranchising some while others remain excluded; the complicity of the speaker in capitalist imperialism, a complicity that his rhetorical practices are designed to veil or obscure' (Hayles 87). It is perhaps no secret that Wiener's cybernetic science was, in the words of Katherine Hayles, 'deeply connected to the military, bound to high technology for its very existence and a virtual icon for capitalism', and that his conception of a merger between human and machines, heretofore known as 'the cyborg was contaminated to the core, making it exquisitely appropriate as a provocation' (159). Such a provocation relates back to revolutionary era America, where 'the dominant theory of the time was that 'black people were not fully human' (Johnson 371). Benjamin Franklin, another one of America's founding fathers, held that the black body was first and foremost to be understood as the property of whites, and therefore any other instituting claim to it was the work of unlawful appropriation (Singh 36). One could argue that the founding of the United States was born of a desire to legalise a criminal enterprise, and in so doing, invert criminality itself so that guilt was cast no longer on the illicit acts of the subject, but now the illicit acts of his objects, and therefore able to totally monopolise the universe of claims.

Nikhil Pal Singh notes that from the founding of the American nation, blackness was seen as synonymous with 'thievery' (36). Carrying out such a logic requires that each citizen take on a sense of individual responsibility for dolling out various forms of punishment in the form of prejudicial thinking, discriminatory treatment, and arbitrary violence against those they perceived as being a threat to the state – at home and abroad. The presumptive context of societal exclusion always bears a dual impression: of both racialising division and civilising aggression, such that Americans perpetually dwell in their national imaginations somewhere between a dominion of capitalism and a theatre of war as the uncanny, parallel universes born of their creation.

During the 1960s and 1970s, this same peculiar institution allowed extrajudicial violence to be enacted against blacks, playing itself out in pacification and counterinsurgency tactics applied in response to both black and third world radicalisms, which once more were met with the great force of American exceptionalism, such that the desire to constrain blackness once again came in the form of wars that grafted the gross inequalities of capitalism onto a premise of racial essentialism: 'the war on crime, the war on drugs, and now the war on terror' (Singh xiii). As economic precarity grows with the progress of capitalism so too does the need for greater and ever more sophisticated forms of securitization, each incarnation of which take on a racialised contour, in the form of racial profiling, immigration means testing, criminalisation and benefits rationalisation, all function to reanimate an awareness of racial propriety as a vital component of safeguarding the national enterprise. Singh's analysis provides a striking genealogy of this perpetual 'war' against blacks within the context of Clausewitz's formulation of it 'as a continuation of politics by other means' and details its symbolic progression as follows:

> By the 1960s, the sense of the intimate proximity of violent racial abjection, of race making and war making overseas, had become integral to black critical discourse…Martin Luther King Jr. observed that the promises of Johnson's Great Society had been shot down in the battlefields of Vietnam, expressing his regret that

"my country is the greatest purveyor of violence in the world today." Twenty years on, when George H.W. Bush was kicking the "Vietnam Syndrome" in Iraq, the rapper Ice Cube tied it the violence of the drug war, memorably describing the Gulf War as a giant "drive by shooting". The acquittal of the New York City police officers who killed an unarmed man, Sean Bell, in a hail of gunfire in 2007 prompted the family's minister to remark, "Here it's just like Iraq, we don't have any protection" (22).

Science, law, and biopolitical regulation have all conspired through these events to conjure a narrative of racial awareness, which at every point coincides with the fortification of violence through its associated, militarized social technologies. In her book *Dark Matters: On the Surveillance of Blackness,* Simone Brown recounts an incident that took place in 2004, when the actor Will Smith was promoting his film "I, Robot", where he plays the police detective Del Spooner, who became an involuntary subject in a cybernetics program for wounded police officers after being in a car accident. As a consequence of his injury, Spooner is fitted with a prosthetic left arm by a company called U.S. Robotics. The same company that created Spooner's arm also has created a legion of robot servants (sometimes referred to in the film as slaves). After his surgery, Spooner is commission by the company to police them. Spooner must use biometric information technology to conduct his duties, to prevent as it were a 'slave revolt'. Spooner, despite the fact that he is a cyborg himself, is portrayed as rabidly 'antirobot' and treats this technological population with grave suspicion throughout much of the film. When Smith was asked by the German press about the effects of the 9/11 terrorist attacks, he reportedly answered,

Whether you're attacked and wounded by a racist cop or attacked by terrorists, excuse me, it makes no difference. In the sixties, blacks were continuously the target of terrorist attacks. Although it was domestic terrorism, terrorism is terrorism. We are used to being attacked. As for a permanent alert, a defensive attitude with which one lives anyway—it has not changed since. No, for me personally, as to my everyday life, the tragedy of

> September 11 changed nothing. I live always a hundred
> percent alert. I was never nervous, anxious, or cautious
> after 9/11 (Smith qtd. in Browne, Dark Matters, 239).

Browne's analysis suggests that Smith needed to exercise another type of caution. Not in terms of being caught off guard by the appearance of racialised terrorism, before and after 9/11, but rather in terms of the censure he would receive from white Americans based on his remarks, which ran counter to their overwhelming desire to distance themselves from the truths that seem so evident to Smith as a black American. Throughout his career, Smith has acted as the (role) model of a generic black masculinity that coerces the appearance of black men to conform with a process of anonymization, wherein their identity is one that is bestowed upon them categorically through a casting system riddled with white supremacist assumptions. It is within that rigid system of apprehension that black actors are given their identities, their schedules controlled by others, and even where they live is dictated by their mostly white industry overseers. In order to be cast, they must have no identifying markers of any kind that might correspond with their previous lifetime as ordinary black Americans. Smith's problem in speaking out is that he has transgressed his commercial agreement to accede to white America's seamlessly constructed image of him and to concede that his every cultural transaction is closely monitored. Essentially Smith is given the choice to either play it safe, or risk commercial boycott of his cultural franchise.

At the beginning of his remarks, Smith speaks of a collective black 'we' that is forced to 'live with a constant feeling of discomfort' and adopt vigilance as second nature to their world view (Smith qtd. in Browne, Dark Matters, 239). By contrast, those who are made comfortable in the face of surveillance technologies and practices experience surveillance as an extension of the white popular imagination. Such surveillance functioned over centuries as a normative state of securitization based on the premise of 'protecting America', and by extension, the planet, from 'Others' who were presumed to be alien to that franchise. The alternative to being included in that franchise was to cast those suspect Others excluded by it, in some sort of policing role in order to unwittingly report on themselves. Therein they would be seen as 'voluntarily' subjecting themselves to measures of control that would be rationalised and conveyed back into the white popular imagination, as speculative proof of their negative conduct requiring of pre-emptive action.

In the 1960s, urban security practices widely adopted the anticipatory techniques of cybernetics developed during World War II by Nobert Wiener, to predict the 'aggressive' behaviour of black populations in urban centres across the United States. Various surveillance techniques were crafted by municipal agencies to act as 'early warning systems', as a means of geographic containment which 'reflected and reinforced racialised geographies' (Young, Pinkerton & Dodds 59). The racialised thinking that gave birth to the United States, also gave rise to an early Silicon Valley whose target was nothing less than projecting a resplendent 'free world,' out from the dark depths of the Cold War. New hybridities were put to the test to fuse education with mobility, militarism with liberalism, and labour with technology in ways that could contain and neutralise the effects of 'anti-capitalist dissent, racialized urban immigration, and grey, industrial density' (McElroy 2). This strategy was as much about the *incitement*, as it was the *excitement*, of consumers who could be driven to participate in their own techno-driven dispossession fuelled by existing economies of urban racialisation and suburban pacification.

What was born of the social laboratory, quickly spilled out to encompass a set of practices fit for comprehensive integration into a greater system of settler colonialism perpetuated through the expanded reach of high tech corporations. While it is possible to say, that theirs was a project of assimilation, it is also possible to assume that a Cold War inspired Silicon Valley, was vastly implicated in the trajectories of race and racism that condition reality to mutate into patterns of virtual differentiation, moulding the experiences of heterogeneous consumers to conform to varying racist, colonial, socialist, and national contexts. Perhaps one can go as far as to say that it could only have been against a backdrop of Cold War geopolitics that the Internet as we know it has materialised, that we who interact with it are still in some senses engaged with a Cold War subjectivity, and that the 'California dream' cannot be understood apart from an index of fears around liberalism and racialisation.

Slave to Technology

A watershed moment for racial encoding happens with the advent of cybernetics in the United States with the publication of Norbert Wiener's seminal volume *The Human Use of Humans* (1950), which is interspersed with racialised analogies having to do with master versus slave and race versus technology. Louis Chude-Sokei argues that this was the case because 'cybernetics as a science of communication and control was explicitly considered in terms of racial politics' (83). The case could be made that Wiener revisits these metaphors of slavery for the benefit of a postwar American and British readership psychically steeped in a rhetoric of resistance related to Nazi Germany's recent maniacal state program of racial domination and forced labour. The racialised politics of this system, after all could not be lost on Wiener, who was a first generation American Jew of German and Polish origin. And yet, this never becomes a part of his intellectual feedback mechanism in the development of his work on cybernetics. Wiener in his book does not refer to race beyond a promotion of the human race and a concern for the arms race. Wiener's commentary while referring to the human race ignores the effects of historically rooted social hierarchies and the struggles of people oppressed by unequal power. Wiener instead chooses to adopt a 'steering' metaphor as a means of manoeuvring culture and economics away from the conduit of understanding power and capital. Therein, cybernetics is not built around a desire to morally reform conditions of labour so that it would not be compelled to accept conditions of servitude, but rather to optimally organise these presumably free bodies to enhance their existing value.

For Wiener, the descent of labour into slavery, corresponds perfectly to fascistic models of economics 'where the arms expansion replaces the growth in consumption and where investment veers from the means of production towards the means of destruction (Deleuze and Guattari 254). However, Hitler's fierce and vehement expression of hatred for Jews and communists, is not his most disturbing feature for Wiener, but rather his ability to wreck chaos on the world at large. In this way Wiener transcodes his dread of National Socialism into cybernetics a life-preserving pursuit, allowing him to mathematically enforce clearly bounded flows of communicative feedback. The numerous classes of bodies that Hitler's ideological programme wrecked were all too human, fleshy and unpredictable, driven by the darkest of drives. Wiener by contrast wanted to introduce a new type of race into the equation by reducing the body to its energies, to fuse their vocation with information, making of them an ultramodern form of capital. Such bodies could withstand 'unresolved discord', translating it in 'semi-permanent and recognisable terms' (Pohl).

Stefan Pohl argues that even as Wiener was ostensibly 'writing to combat what he discerned as the deadly freeze-framings of fascism', he was also inviting 'heterogeneous beings' to now appear and interact instead, 'ecologically, as if environments, as if spatially, temporarily outside' of 'the reductive and homogenizing violence of fascism,' and thus somehow beyond the reach of the old sciences of war (Pohl). Nonetheless, the new science of cybernetics, could not help but rely on an 'ever-widening interface, between command, control, and communication processes' situated within 'a diverse array of machinic, biological, and social systems' in order to produce coercive applications for the leading figures of post-war social science – 'Margaret Mead, Talcott Parsons, Gregory Bateson, Kurt Lewin, and Robert K. Merton, among others' (Pohl).

Wiener for his part naturalised this joining of technology to the body, through the existing conduits of the social and cultural sciences of the time including sociology, anthropology and psychology. These sciences all had a hand in the formation of racial science in the closing decades of the nineteenth century and opening decades of the twentieth century. All of the imagery and meaning that was formerly assigned to the differentiation of racial groups is now made constructive towards a project of hybridisation, a mixing as it were not of the races, but of the corporation with the self-fashioning of humankind. Within this environment, race can only be defined through set parameters that reaffirm a conformity to type and creativity that stays within the bounds of the mainstream. With few possibilities for autonomy, a new kind of slavery emerges born of an automated merger between connectivity and co-dependency through the rapid assault of sense-making screens, that increasingly spare individuals from the burden of thinking, subjecting them to newer forms of mastery in the form of remote management techniques, that provide them with an illusion of self-governing, all the while ensuring their class immobility and societal abandonment.

Corporatist cyborgs cannot invent themselves any more than their mercantilist slave counterpart could, their designation is proprietary, in its every move, case, utterance and iteration their value is acquired by dispossession. Stefan Pohl maintains that capitalism, in its 'ultramodern or cybernetic mode' is bent on 'incorporating the entire world as its parasitic playground for profit.' Furthermore, through 'cybernetic control practices' bodies are guided towards the goal of their 'material survival,' which can only be achieved through their narrow advancement 'within the [strict] confines of a cruel, complex, and contradictory socio-economic system' (Pohl).

As a child, 'Norbert Weiner, the pioneer of cybernetics, dreamed of becoming a naturalist and explorer,' as an adult he would have a different task, that of 'curating and reordering a world now fully revealed to itself... a world that, by the mid Twentieth Century, was already widely regarded as saturated with data' (Duffield 1). The world of personal experience, documentation and indexing was extinguished in the twin conflagrations of World War I and World War II, allowing new orders of experience to emerge 'relating to the structure, organisation and growth of information. The birth of cybernetics began the search for technology invested in communication, self-referentiality and prediction' (Duffield 1). Within this new context, race must become something that is emergent; a category that must be secured through analysis of the empirical behaviour and livelihood choices of those inhabiting it. Those enveloped by race, were no longer to be the objects of projection, forced to act as it were as the grid lines for exploration and conquest. Their enclosure, would instead take place at the intersection of anticipation, vertical surveillance and intervention from within a ubiquitous electronic atmosphere, where informationalised bodies could now be remotely mastered by corporations rather than states.

Within this economy the cyborg, i.e. the labouring body forced to hybridise with technology, appears as a way to give a face to the recombinant processes that translate racism and colonialism into a postwar racialised architecture conversant with an informatics of domination. The blueprint for this new architecture was conceived in a Cold War Era Silicon Valley. This region would provide the ideal conditions for a cybernetic infrastructure to materialise through the fusion of militarisation and education, engineering a world perpetuated on the myth of its own universalism. The designated outcome of the Cold War was to re-engineer the social and political minds of individuals to be receptive to 'new notions of a universal humanism' (McElroy 2). The new mind set would be predicated on an acceptance of science and technology as the arbiter of essential gateways through which individuals could maximise personal freedoms. Militarily employed scientists and psychologists would become the societal experts charged with culturally legitimating this mind set beginning in the late 1940s. The freedoms promoted from its inception were differentially coded according to somatic and geographic boundaries which carefully separated the aspirations of white suburbanites from their racialised urban counterparts. Though separate, each imaginary held within it a principle desire for security and containment as presumably 'democratic' features of spatial management.

In his memoir *Ex-Prodigy* (1953), Wiener 'made a statement about diagrams that also imagined a new future into being, that bears on our contemporary concerns with ubiquitous computing, data and visualization' (Halpern 12). Wiener writes of how he longed to be a naturalist, '"as other boys longed to be policemen and locomotive engineers,"' but that this path was darkened to him by his dim awareness of the fact that '"the age of the naturalist and explorer was running out, leaving the mere tasks of gleaning to the next generation"' (Wiener qtd. in Halpern 12). Wiener's age of discovery would be narrated not by a desire to grid life itself, i.e. through the classificatory sciences of zoology and botany, but rather through the diagramming of complicated structures and the charting of problems of growth and organization. Complexity would emerge as Wiener's field of conquest and his mode of interrogation, in a Cold War context that was now limited to organization, as opposed to exploration, as had been the case during his US military work during World War II. In his memoir, Orit Halpern observes, 'Wiener indicated a desire to see an older archival order, adjoined to modern interests in taxonomy and ontology [cultural, biological], rendered obsolete by another mode of thought invested in prediction, self-referentiality, and communication' [formation and mediation] (my bracketing)' (12). If this is true, as Halpern concludes, that 'Wiener dreamed of a world where there is no "unknown" left to discover, only an accumulation of records that must be recombined, analyzed and processed,' it could equally be the case that Wiener believed this could only be achieved through the capitalisation of science (Halpern 12).

In a world where there is no unknown to discover, it would be an error to underestimate the use of information as a weapon within the superstructure of a greater economy of complexity. While Wiener's cybernetic order may 'brush aside a number of outmoded concepts, such as creativity and genius, eternal value and mystery', concepts whose uncontrolled (and at present almost uncontrollable) application,' still 'would lead to a processing of data in the Fascist sense,' because its perceptual terms continued to rely on a "sense of the universal equality of things"' (Benjamin "Illuminations" 218). This was because the goal of record keeping, of data analysis, is nothing less than a mechanism, 'to pry an object from its shell, to destroy its aura, and to extract value from its unique marking by means of its capacity for reproduction (Benjamin 223). Thus data 'is manifested in the field of perception', through 'what in the theoretical sphere is noticeable in the increasing importance of statistics' (223). Cybernetics, therefore, acts in its own way, as a proto-fascistic field of force, is concerned with 'the adjustment of reality to the masses and of the masses to reality is a process of unlimited scope, as much for thinking as for perception' (223).

When considering Walter Benjamin's conclusion about the potential for information to act as a handmaiden to conquest and genocide, it is useful to recall IBM's strategic alliance with Nazi Germany -- beginning in 1933, which would commence in the first weeks that Hitler came to power and continue throughout World War II. Edwin Black's landmark volume, *IBM and the Holocaust: The Strategic Alliance between Nazi Germany and America's Most Powerful Corporation* documents IBM's identification and cataloguing technologies of the 1930s and 1940s, which efficiently targeted Jews, making it possible for them to become progressively subject to 'asset confiscation, ghettoization, deportation, enslaved labour, and, ultimately, annihilation' (8). This was a monumental feat of organisation that applied automation to persecution and in so doing generated a new reality in Germany constructed from data. IBM not only was able to achieve the identification of the Jews through censuses, registrations, and ancestral tracing programmes, but equally to run railroads and organise concentration camp slave labour all through their determination to design complex solutions responsive to the Reich's immediate systemic requirements to create a near perfect feedback loop between corporation and consumer. IBM's Hollerith 'codes compilations, and rapid sorted had enabled the Nazi Reich to make an unprecedented leap from individual destruction to something on a much larger scale' (Black 365). It was in fact, the 'data-diven denouement of Europe's Jews courtesy of IBM's technology that afforded Hitler the ability 'not just to kill Jewish people, but to exterminate Jewish *populations*' (365). Therefore, it is no exaggeration to say that the company who had billed itself on "solutions" based thinking, responded through its singular statistical means to furnish the implementation of "The Final Solution."

Throughout modern history, bodies that make up these disaggregated masses are always already '"cybertyped", branded with a certain set of affective identity markers' (Galloway "The Interface Effect" 139). Whenever a body distinguishes itself, it always already articulates itself 'as a body codified with an affective identity (gendered, ethnically typed, and so on), determined as such by various infra- structures both of and for identity formation' (139). The complexity involved arises from not simply from the fact that bodies must always articulate themselves, but that they must do so categorically if they are to be intelligible. It is not always the case that bodies that stand out are marked for censure or elimination, in many instances in fact it is 'their difference is mobilized as fuel for value creation in the marketplace' and therein value becomes something that wholly involuntary as labour (139). Therefore, differentiation is no longer actual, it is technical, and where differences arise they become the stuff of algorithmic record and the human part of an identity politics that relies upon the subtractive.

The virtual universe summoned into being by cybernetics is only superficially concerned with freedom. Within its underside lies the capacity for an endless debasement of the self to commercial exploitation predicated on market accumulation. The masses if they can be said to exist at all in this new world order, subsist along lines of radical neutrality. Within such a space, Alexander R. Galloway, believes the trick to the survival of the individual is to 'abstain from the assignation of traits, to abstain from the system of biopolitical predication, to abstain from the bagging and tagging of bodies' (140). We must resist the desire to accede to being tailored and targeted through our affective predilections, and to the transaction of likeness that demands in return that we are limited to the horizon of racialised identities that have their founding in a former liberal political subjectivity we no longer have the luxury of inhabiting as a chrysalis of implied justice and assumed neutrality, rational conduct, and so on. Galloway implores us to 'recall the conceit of white privilege: to cast off the fetters of race' without regard for 'those bodies of color for whom this is not an option' or have 'no desire to abandon themselves, to abandon their culture, to abandon their history' (141). The task ahead must be for all of us evade the capture and reproduction of ourselves as aggregate consumers, differential from one another solely at the level of affective behaviour and biometric identification, and achieve a form of subjective autonomy that resists complicity with the symbolic violence of digital media, through a profound sense of recognition that its technological ethos traces back its lineage to the material history of race.

Throughout that history, including in its latest virtual incarnations, the material lives of African Americans have remained overwhelmingly disadvantaged, precarious, and vulnerable to what has come in to stand for American culture as a whole; at once a stalwart of celebrity and charisma and a prison for information and assembly. America's racial nomos has been individualised and encoded with practices of role-modelling and mentoring which favour an understanding of success built on a false liberation from past injuries and insults, and a faux autonomy based on entrepreneurial acumen, all while retaining the message that one's own life is a commercial operation. This neoliberal narrative of racial uplift for African American's pointedly leaves out one glaring feature; than an explicit appetite for market domination and the manipulation of profit is what motivated their former subjugation as a collective body. This deeply entrenched racial hierarchy has been retooled into 'a programme of hi-tech manhunting' set to 'encompass further struggles, both explicit and inferential' (Gilroy, "Race and Racism in 'The Age of Obama,'" 8). This older operational manual underwrites 'the significance of an emergent racial hierarchy (that is often mapped onto differences of faith and language) and its relationship to capitalism, to global inequality and to the prosecution of an apparently endless war on terror' and domestic extremism (8).

In the final decade of the twentieth century race disappeared from public and government discourses. Its erasure coincided with the formation of policies regarding Internet infrastructures and access (Nakamura 202). Lisa Nakamura pinpoints this as a specific historical moment when neoliberalism materialised as a global hegemonic force, capable of producing a particular set of racial formations that were attentive to the rapid expansion of virtual interfaces and networks (202). Rather than innovate a conception of race, 'the massification of the Internet, has relied upon traditionally racialized representations of people of color' to promote itself as an unprecedented tool of information (208). Nakamura observes that 'on the Internet, users *do* as well as *are* their race' and moreover, 'this networked racial positioning broadcasts this *doing* in ways that explicitly un-"cover" race' (206 emphasis mine). This epistemological 'discovery' of race is further enhanced through digital systems such as facial recognition software that operationalise and instrumentalise race. These systems also generate new containers for its apprehension and assignment by giving a precise location to race as an essential trait. The consumer is therefore given a second skin in the form of an online identity that must further reified and essentialise white privilege, as well as the discrimination of blackness, meaning that 'race' is now sourced through customised identity management systems that operate on a micropolitical level. These new platforms seek to universalise race while still using it as a tool of justified, or valorised elimination, as yet another of Silicon Valley's assimilatory projects of human engineering and defensive modernisation in a cybernetic world of human governance.

Chapter 4
Big Blue: The Anthropocene
and the Advent of Planetary Finitude

"Forgetting extermination is part of extermination." — Jean Baudrillard, *Simulacra and Simulation*

"Today, the solitary inventor, tinkering in his shop, has been overshadowed by task forces of scientists in laboratories and testing fields. In the same fashion, the free university, historically the fountainhead of free ideas and scientific discovery, has experienced a revolution in the conduct of research. Partly because of the huge costs involved, a government contract becomes virtually a substitute for intellectual curiosity. For every old blackboard there are now hundreds of new electronic computers" - Military-Industrial Complex Speech, Dwight D. Eisenhower, 1961

We are all haunted by the possibility that there may be some hope for the future. -Ernst Jünger

Atomic Universalism

The end of the future coincides with 'the instant the U.S. military exploded a nuclear device at the Trinity site in New Mexico… at 05:29:21 Mountain War Time (+2s) on July 16, 1945' (Mirzoeff 140). This is the moment when our blue planet was reconfigured to obey the coordinates of an emerging technology culture whose geopolitical imprint specified a necessity for life on Earth to mobilise in response to what was deemed its "Great Acceleration", its forward thrust towards its own potential annihilation, from the bankrupting of its natural resources on one hand, and on the other, the fragmentation of life itself to the ontological status of building blocks and components. This would be the era where life and its artificial counterpart would begin the long course of their corporate merger in earnest. It is also when life would submit to a new radical manoeuvring, through a globally-orchestrated technic of politics and economics, bent on striking a balance between the emergence of planetary-scale technology and micro-scale technology through a variety of experiments related in one and another way to social engineering. A new sense of reality itself was effectuated by subtle manipulations of algorithmic code, and discipline of our imaginations through cybernetics, allowing for new epistemologies to emerge. Through them, humanity began to conceive of our Earth as a technosphere and our mission to mine its resources in order to platform new worlds to service our material dependencies to come. With the advent of this new age steeped in fascination for the atomic, digital, molecular, and inorganic, humanity came to fully conceive of itself as the agent of its own operational destruction.

The activation of this pattern of thought permitted a type of humanistic liberation theology to come to the fore that swept its way through not only the intellectual domains of research and development, but equally philosophy and aesthetics, all of whom were eager to take their place as actors in what was most self-consciously deemed a revolutionary era for humankind. And so, it was that we entered unwittingly into what we now commonly refer to as "the Anthropocene". We now readily engage with the performative elements of this terminology whose advent neatly coincides with the historical invention of "the environment" as Earth's global operating system, the elevation of art and culture as mediums of calculated experimentation, and the integration of Silicon Valley as the global nexus of a rapidly ascendant military-academic-industrial-complex. We dwell now on a planet littered with the remains of these largely destructive postwar conceits, taking their form variously through the appearance of ubiquitous petrochemical waste, the wholesale corrupting of nature through deregulated international trade and foreign investment, the advent of new forms of race, class, and religious disaggregation, the monetisation of information, and the mass replication and storage of data.

The dawn of the geological epoch of the Anthropocene is widely held by scientists to have commenced somewhere between the mid-1940s and mid-1950s. Its advent is closely associated with the start of the nuclear age, and marked by the appearance of radioactive waste material for the first time in the Earth's geological record. In that sense, the Anthropocene can be conceived of as a kind of second planetary big bang. Such events that are, by definition, so globally destructive as to trigger the founding of a new geological epoch. These types of events send out a 'signal' through deposits made into the Earth's geological record. Damian Carrington gives the example of the extinction of the dinosaurs 66 million years ago at the end of the Cretaceous epoch. This epoch 'is defined by a "golden spike" in sediments around the world of the metal iridium, which was dispersed from the meteorite that collided with Earth to end the dinosaur age' (Carrington). In terms of the Anthropocene, its golden spike is believed to be the 'radioactive elements from nuclear bomb tests, which were blown into the stratosphere before settling down to Earth' (Carrington).

Other signals were being recorded around this time to indicate the dawn of a new geological epoch through the appearance of 'plastic pollution, aluminium and concrete particles, and high levels of nitrogen and phosphate in soils, derived from artificial fertilisers' (Carrington). Together, all of these elements appear to closely sequence with the apex of 'The American century'. This term was coined by *Time* magazine publisher Henry Luce on February 17, 1941, was intended to signal the fact that history had conferred global leadership on America. However, unlike its imperial predecessors, in Luce's estimation 'the power conferred was global and universal, rather than territorially specific' (Harvey 50). David Harvey observes that the definition allowed American power to take on an epochal rather than imperial mantle on the world stage. This definition of power, not only distances America from Britain but also marks an end to its greatness through a "golden spike" applied directly to the historical record, which at once mars and blurs the recording of the geographical signal of empire in the centuries proceeding America's fundamental break with it. Whereas empires are open to challenge through their rising and falling, Luce's American Century 'suggests an inevitable destiny' (Harvey 51). Therein 'US global dominance was presented as *the natural result* of historical progress, implicitly the pinnacle of European civilization, rather than the competitive outcome of political-economic power' (51, my emphasis).

Luce's essay bears out this argument by giving the examples of the global preponderance of 'American jazz, Hollywood movies, American slang, American machines and patented products', as a critical mass of material articles 'that every community in the world, from Zanzibar to Hamburg, recognizes in common' (169). This commercial phenomenon signalled to the world that America 'was beyond geography' (Harvey 51). That said, Luce believed that America's enormous scope of influence on the world was realised in one sense, 'blindly, unintentionally, accidentally and really in spite of ourselves' (169). As a consequence of this superficial recognition, the power that America wields prior to World War II, is in Luce's words, 'trivial' and 'very human' (169). Luce concedes however that 'there is a great deal more than that,' insofar as America was 'already 'the intellectual, scientific and artistic capital of the world' prior to formally entering World War II (169). This paradox of American's being construed as both intensely naïve and radically sophisticated would come to permeate the global imagination and in doing so positively bias the calculation of 'the ultimate intelligence and ultimate strength of the whole American people' (169). What Luce's essay anticipates is how the combined achievements of America's core 'engineers, scientists, doctors, movie men, makers of entertainment, developers of airlines, builders of roads, teachers, [and] educators' would persuade the world of America's superiority of 'imagination' and the supremacy of its 'enterprise' in the immediate decades following World War II (170).

What Luce failed to recognise at the time of his essay writing, was how entry into the Second World War would cast America not merely as 'senior partner' to Great Britain 'in any sort of partnership with the British Empire' but as its Cold War superior due to its nuclear capabilities and their planetary repercussion (164). As it turns out, what the new America Century was not beyond was geology. Thus, it would seem 'empire is not bound by the territories under direct metropolitan administration but extends to all spaces enmeshed in regimes of indirect control and subject to the disciplines and magnetic pull of imperial cores' (Golub 66). Perhaps this instant can also be said to refer to the moment in which colonial and imperial ambition formally switched over from the British Empire to the

American one. With this transfer of power from the petrochemical to the nuclear option, emerges the scenario of the 'end of the world'. Here resources are controlled by those with the greatest access to power. In a world seemingly teeming with abundance, it then becomes possible to imagine the goal of humans being abandoned by their old orders in the wake of these new technologies and for humanity itself to be forced to embrace knowledge of life on earth as something inevitably dwelling at the threshold of annihilation. It is here, in this world, that we must survive in the wake of the robustness of the ultimate power that now reigns here.

The Cybernetic Technosphere

In Santa Monica, California on 14 May 1948, 'the study of the future began at the RAND Corporation', an independent think tank set up to 'undertake research and development (R-A-N-D) for the military' (Beck 37). The RAND Corporation was originally tasked with orchestrating America's Cold War through the material development of nuclear and aerospace weaponry, and perhaps more significantly, the cerebral cultivation of its ideology of counterinsurgency and theatre limited proxy warfare. Through these efforts, California emerged as the nexus of the military-industrial complex. California was cast as the cutting edge of Western civilisation, a place powered by a potent combination of massive military spending and unrivalled research and development. A "golden spike" in annual investment by the federal government triggered its economy to multiply exponentially during the years of America's involvement in World War II. By 1945, annual spending by the federal government in California had reached a historical zenith of '$8.5 billion,' stimulating a merger of the outputs of military contractors, universities, and martial metropolises into an economic configuration of infinite flexibility (Gilbert 28).

The Cold War was very much a product of this overheated spending environment where the objective was to keep defence spending high to stimulate economic growth in other closely related sectors of the State, such as filmmaking and manufacturing, that together would perpetuate California's economic ascendency. The Cold War, from its inception, coupled the nuclear arms race with the space race, which acted as a signalling point for neo-colonisation projects both of a terrestrial and extra-terrestrial nature. The capability of these projects relied on the bodily toil of a legion of migrant workers of colour drawn to California for its abundant manual work opportunities, while the leading-edge work went to an elite core of white male venture capitalists, entrepreneurs, and engineers. Everything from agriculture, aeronautics, entertainment, and universities followed this pattern of industry in the sunshine state.

Los Angeles, San Francisco and San Diego figured as the trifecta of a plan of growth that united military investment with firstly urban expansion, and subsequently suburban expansion. These areas provided the necessary population to perpetuate the 'synergistic core of the California economy: close cooperation between the military and civilian communities' (Gilbert 29-30). The superior beneficiaries of that arrangement have been the major regional universities and Silicon Valley, who have spurred a class of scientific-technological elites to become the thought leaders shaping American public policy. Their thinking has often been comprised of a rough translation of war-research into the civilian arena as a means of facilitating a new economy where academic, military and industrial concerns seamlessly merge. This merger began in earnest during World War II, when 'both the United States military and its civilian defense industry partners came to acknowledge the enormous complexity that had come to characterize both modern industrial planning and national security policy-making' (Gilbert 33). The American government quickly established operation research units comprised of research teams of physicists, mathematicians, engineers, aero-dynamicists, chemists, economists and psychologists, all united in the task of predicting the challenges and solution appropriate to managing a postwar world.

The most celebrated of these was the product of a military-corporate merger between the air force and Douglas Aircraft Company of Santa Monica - the RAND Corporation. 'RAND's early work secured the public welfare largely by providing the brainpower for determining what would happen in the event of a nuclear war with the Soviet Union' (Gilbert 33). Their collective mindset was always directed towards the furtherance of "thinking the unthinkable" through the lens of anticipating a world of perpetual conflict. What they applied to war-making was an ethos of corporate managerialism, wherein its '"systems analysis" approach, moreover, promised cost-effectiveness and conveniently, if dangerously, reduced much of the complexity of postwar political struggles into a form digestible to policy-makers' (Gilbert 33). RAND relied heavily on cybernetic research to fortify its authoritative, and indeed, authoritarian approach to security matters, applying systems thinking and quantitative rational management to each seemingly intractable problem, becoming effectively the social, political, and economic architects of America's early nuclear age.

At the same time, 'cybernetics was also supposed to transform the economy into an ecology of feedback loops in order to control social unrest and potential revolutions' (Pasquellini 318). The body and the economy were therefore made to operate in a delicate relay, wherein information became the regulatory mechanism of good governance. 'More exactly, cybernetics was the normative project of power in the age of information machines ... (318-319). Within such a scenario it is not just the individual human body that must be interpolated in the civilian imagination as a vast operating system holding within its apparent bounds a formidable cosmology, but also the notion of Earth as a metabolic entity, operating through the perpetual networking of energy and information, resource and extraction, data and mining which together formed the prototype of the age of the Anthropocene, in which such infrastructural colonial features are now more readily integrated and computed than ever before in our human history.

Today the metabolism of the planet has increased dramatically, we move within the economy of transparency, rather than dwell within a nature of secrecy. 'The nature of secrecy changed...largely as the result of the dematerialisation of data flows' (Trotter 15). 'Thereafter, the very telecommunications technologies which were in the process of vastly extending the secretive, quasi-global reach of commercial, political and military power became the most reliable means of its exposure. Knowledge no longer existed apart from networks; and networks are vulnerable in ways that the individual courier isn't' (15). We now communicate solely in mass thanks largely to an imperial ambition seldom recalled when we think of the advent of our modern wireless age: The Spanish-American War of 1898.

Establishing America's Communication Hegemony

America's war with Spain was brief but had lastly impact insofar as it resulted in its first acquisition of lands beyond North America: The Philippines, Guam and Puerto Rico. Soon after, the United States acquired the eastern islands of Samoa, annexed Hawaii and the Panama Canal Zone, and established Cuba as a protectorate. The American political body thereafter expanded in unprecedented ways that required that the burgeoning nation's imperial ambition now advance towards technological domination. The battlefront, in this case, would be the control of telecommunications within America's new sphere of colonial influence. 'The cutting-edge technology of the time was transoceanic cable, and two-thirds of the cables that existed were owned at that time by Britain. Effectively, Britain during this era was the world's global telecommunications hegemon, and the United States' greatest imperial rival' (Adams 334). Prior to this moment Britain as the world's leading imperial power had held a monopoly on information technology dating back centuries, as it had also done on militarised technology at sea. The United States Navy, therefore, became the most obvious agent through which the breakthrough both technological strongholds.

At the time, 'wireless firms were depending on government contracts for their commercial survival and therein their interests closely aligned with the imperial ambitions of their best customers ... the U.S. Navy' (Adams 334). The joining of these commercial and military interests acted as a means through which not only to disrupt Britain's imperial lines in what we soon thereafter be called 'the American Century', but eventually to tilt the global nexus of power irrevocably westward onto an alternative frontier of technological innovation. Strategic telecommunications would no longer be realised as the product of Britain's imperial legacy on the colonial East Coast, but rather as the bequest of the Spanish on the imperial West Coast, 'and in so doing gave impetus to the creation of America's modern military-industrial complex with the dawn of the twentieth century' (334). Within sixty years of that war, California would become the leading state in the United States for defence contracting. By the end of World War II, San Francisco would boast that it had more government offices than any city apart from Washington, D.C. The Bay Area thereafter functioned as a key region of strategic importance to the advancement of the entire nation.

Ideological Telegraphy

This uniting of the American Coasts takes place through another technology predating the advent of transoceanic cable in the first half of the nineteenth century. Pamela K. Gilbert credits the early nineteenth century with definitively sourcing 'consciousness in the nervous system and its distributed terminals on the body's surface' (2). Paul Gilmore asserts that telegraph emerged as a pivotal technology in those same years through its ability to demonstrate to the white masses the power to release thought from the body. In so doing it prompted whites to consider their ability, through the aid of technology, to assert greater and greater dominion over both the boundaries of the natural world, as well as the physical bodies of others. Throughout the 1830s and 1840s, 'the telegraph was described as 'enabling a disembodied Euro-American mind both to conquer the natural power of electricity and to enslave the bodies of blacks and other "tropical" races' (Gilmore 810-811). The strategic use of energy, to overpower America's others through technology was never solely a visceral ambition. More so it reflected a desire for the nation and its peoples to accede to profound disembodiment with the exception of their brains in order to 'become master of the physical features (and implicitly the peoples) of the Orient, Africa, and Asia' (811). 'As an anonymous poem in *The Atlantic* phrased it in celebrating the first transatlantic cable, "Through Orient seas, o'er Afric plain, / And Asian mountains borne, / The vigor of the Northern brain / Shall nerve the world outworn"' (811) The European brain as an organic compound alone cannot accomplish this feat, but rather through the application of electricity will gain the capacity to reinvigorate itself. It will do so through the medium of technology through the telegraph and the transoceanic cable, each derivation allowing it to communicate itself with both the spiritual and the physical realm. Through this new adapted nature over time there will be no further requirement for individual white bodies to penetrate into nature to achieve conquest. Rather through the enhancement of the Euro-American brain, it will be possible to construct a unified white national body and for America to ultimately emerge as the world's nerve centre.

During the 1850s and 1860s, China would come to grapple with the British introduction of telegraphy into its national interior. It would be marketed to the Chinese as a tool of commercial enterprise, but what Chinese officials were quick to realise was its potential to introduce 'irreversible and foreign-dominated change' into the nation (Bruun 41). Many Chinese reformers and intellectuals sensed this threat and attempted to counter such ambitions through an internal mobilisation of the native population in the cause of anti-foreign resistance. Fengshui was exercised as a conduit of such resistance because it figured as an essential component of the ancient science of geomancy; the practice of which remained wildly popular amongst China's general population. Ole Bruun observes that interest in Fengshui increased in proportion to the influx of Western missionaries and European entrepreneurs into China in the closing decades of the nineteenth century. In those years, Fengshui was afforded contemporary significance by the Chinese government who cleverly reconstructed it as a tool of foreign relations. Its centuries old technology based on terrestrial magnetism was redeployed to act 'as a buffer mechanism to ward off foreign expansion' (47). The Chinese government heralded their defence of Feng shui 'in a democratic oratory alluding to the undesirability of changing the faith of the people' (47). Fengshui emerged as a potent anti-colonial technology in these years granting the Chinese government a means through which to ward off the influence of both Western religious and commercial entities determined to refigure the Asian world in accordance within their liberal democratic values. During the subsequent two-year period of the Boxer Rebellion in China (1899-1901), Boxers attacked both telegraph and railway lines as a means of direct action against Western imperialism. During the Rebellion, 'telegraphic access to and control of information became vital in surviving the complex historical drama involving court politics, the xenophobic Boxers, the important provincial governors, and agitated foreigners' through the popular uprising (Zhou 76).

Ultimately, it would be the telegraph that would be instrumental in putting down the Rebellion. That action would specifically come at the hands of Sheng Xuanhuai the then director of the imperial railway and telegraph systems. Sheng was at first a supporter of the Boxers, but would subsequently decide that they presented an imminent danger to Southeast China. He eventually used the telegraph as a means of conducting political exchanges and in this sense, became a 'nexus of information from whom local governors sought information about the Boxer Rebellion' (Zhou 76). Sheng used his person as an instrument as director of the Imperial Telegraph Administration to manipulate the rely of 'political information by selectively blocking or speeding up communications through coded telegrams, in sharp contrast to circular telegrams' (Zhou 74). Such circulation implied a democratic function; the telegraph, as it were, adopting a public voice. The coded telegraph, by contrast, was by definition an exclusive medium that required an understanding of an elite discourse to decode. Whereas one implied and fostered collectivity, the other implied and fostered selectivity. This pattern of communication would carry significance as a tool of political expediency in China right up until the founding of The People's Republic of China in 1949 when a new era of Chinese politics would commence.

Yongming Zhou discerns that 'from the very start, the telegram was put under rigid government control' (Zhou 131-132). The arrival of the Internet into contemporary Chinese society can be very much seen through the prism of China's previous resistance to the arrival of the West's technological communications. The internet was a tool of imperialism once again poised to revolutionise the Asian world by directing it into faith in western liberal-democratic society and free market capitalism. Once more China's senior officials responded by installing localised breaks and codes to that progression of Western technology to preserve their distinct form of cultural nationalism. The Chinese government continues to make a concerted effort in partnership with private capital, to cultivate a distinct internet economy in China. This is evident through state support of the online commercial platforms Baidu, Alibaba, and Tencent, which were specifically financed and assembled by the Chinese government to counter the commercial and ideological incursion of their US equivalents Google, Facebook, and Amazon in the Chinese communications marketplace.

Climatic Conflagration

In addition to the aggressive promotion of new communications technology, there was a climatic component to Western attempts to colonise China at the turn of the twentieth century. In Mike Davis' analysis in his seminal text *Late Victorian Holocausts* he concludes that the period of greatest intensity for various European incursion into China's interior 'fatefully coincided with synchronized global drought, which we now know was caused by back-to-back El Nino events in 1896-97, 1899 1900 and 1902' (4). Davis characterises the period 1896 to 1902 as 'an epochal crossroads where fused social and environmental cataclysm completed the imperialist division of humanity into two worlds' (4). The year 1900 marked the peak of this cycle of drought, flooding and famine. These catastrophic natural events were attributed by ordinary people to the construction of a preponderance of foreign missions, churches, and cathedrals in the Chinese countryside. Locals believed that this construction had disrupted 'the fengshui or geomantic balance of nature, thus awakening the Earth Dragon' and causing climatic disaster (179).

That year's drought was followed by a devastating flood that led to mass casualties and starvation amongst the survivors. The hunger, disease, and cold was made all the more bitter by the near-universal belief amongst the people that the disaster was of foreign origin and that in the midst of Chinese suffering the great powers of the West, including Britain and Germany, were dismantling China's sovereignty piece by piece, and in the process destroying the imperial foundations on which the people had relied upon for relief in such dire circumstances. A resistance movement gradually emerged, which became known throughout the West as the 'Boxer Rebellion', called so because the English referred to the Chinese tradition of martial arts as 'Chinese Boxing'. This was largely a working class and agrarian movement who targeted Christian villages and local authorities to provide a potent combination of para-governmental defence and predatory theft that appealed to their vulnerable and starving neighbours in greater and greater numbers as the environmental crisis idled their normal labour activities.

Their numbers would eventually grow to the tens of thousands as Chinese civil society further deteriorated. David J. Silbey argues that, 'in a sense, the Boxer movement was deeply conservative, concerned with ensuring the triumph of tradition over progress; the inherent contradiction of the movement was that they aimed to enforce tradition in a radical way, one that disrupted and overturned the central and dominant structures of Chinese society and government' (67). The Boxers were an anti-feudal movement aimed at decentralising long existing power structures within China and more recent foreign commercial and religious occupations of its landmass. In order to materially enact their campaign of uprising against these constituencies, 'individual groups of Boxers acted largely without regard to other Boxer groups, but with the same consistency of aim' (Silbey 67) The Boxer movement was a product not of unified leadership nor communications, but rather of a code of message and method followed seemingly spontaneously by their multitude of affiliates.

Silbey contends, 'we should not mistake that larger decentralization for weakness or a lack in intelligence' (67). Rather, this lack of centralisation was used strategically both to outwit government and foreign agency. Most significantly, their lack of structure from a martial perspective made it difficult for authorities to specifically target them nor anticipate the pattern of malleable strike action. Their recourse to 'headlessness', therefore, was a strategy of both physical dispersal and counter-intelligence. Despite their eventual overwhelming defeat by a military cadre of Western powers, what the Boxers nonetheless succeeded in doing was to establish China's first momentous effort at counterinsurgency against the West in the twentieth century. Mao and his people's revolution followers would later study and adapt the Boxer's tactics of rebellion to wage their own sophisticated campaign of resistance to the West within an ongoing territory of colonial warfare.

Initially, it was a combination of smallpox, malaria, bubonic plague, cholera, and dysentery that fundamentally weakened the Chinese people, making them subject to a campaign of foreign invasion that had to be countered by the remaining population who literally could stand up against them. These desiccated bodies denied food and water by the failure of the monsoons, faced exploitation as a matter of white opportunism. Asset stripping China of its communal lands and harvesting its remaining labour strength in mine and plantations, no doubt in exchange for food, allowed empire to flourish essentially from famine. Tens of millions of rural poor people died in appalling conditions while their European betters looked on in horror, but offered no assistance. They indeed exacerbated the suffering by expropriating the products and labour of the Chinese for the sheer benefit of European economies. Mike Davis argues it is this behaviour which specifically led to the drastic decline of 'state capacity and popular welfare for famine relief, that seemed to coincide lockstep with the empire's forced "opening to modernity" by Britain and other powers' (9).

That same modernity would attempt to force itself on China infrastructurally and politically for the next 50 years, with the Boxer Rebellion seen as the last of the resistance movements to that progress, felled as it were by a potent combination of drought and murderous violence. One begets the other in this scenario, and yet these were also the products of a liberal approach to capitalism that made it both free and binding towards these Asian nations whose native populations faced as much as anything an onslaught of western technology that when supplanted wholesale onto their shores was profoundly disruptive and ultimately deadly. This onslaught decimated thousands of years of deeply rooted cultural and market institutions. Essentially what the Europeans were bringing in was an operating system wholly incompatible with local codes of societal conduct. These changes decimated community structures and in doing so directly lead to the catastrophic death of millions. As Mike Davis succinctly puts it to the Europeans 'this was merely "a systems error"; for the native population, it was a forced error that few of their economies would ever recover from' (10). It was, in fact, a deadly virus, a bioweapon used to wipe out the local economy in favour of rapid capitalistic formation.

This feature of violence extends to this century and its aggressive insistence by technology companies that it must 'open' China in order to expand their now saturated European markets. Colonialism is always the product of such saturation, as is the need to exploit local labour to secure the progress of this territorial invasion into local economies. This is imperial self-interest at play and yet the argument is always one that it is in the interest of the local people to be brought into possession of the West's superior systems of governance. Demise comes in many forms and the violence exacted by markets operates on registers both fast and slow. Ultimately, the rate is determined by global economic shock waves, that may take years to show up on the other side of the world. So, it was in the late nineteenth century, so it is today in the early twenty-first.

Once again climate change and economic processes together steer the world towards this flawed technological solutionism that has been the hallmark of colonialist ambition. It is worth also remember the hand state criminality, terrorism and war plays in priming these markets for eventual invasion. Natural disaster and epidemic disease allow for these markets to become further weakened into a position of sovereign vulnerability. It is not an exaggeration to say that poverty at this national level had to be invented by European liberal market capitalism. It was engineered alongside the advent of new technologies to both create European modernity and to systematically justify the necessary subordination of its others. If one wants to trace the path of its "new world order" one only need follow the bodies of the poor who have paid with their lives for it.

What characterised Boxer militancy was its potent combination of martial arts fighting techniques 'with spirit possession and invulnerability rituals derived from the underground White Lotus sect' (Davis 183). Beyond rousing their spirits and allaying their anxieties about their power to ward off foreign enemies, the Boxers filled their bodies with food. Sustenance was obtained from more well off Chinese neighbours, but perhaps more significantly the food stores of Christian villages and missions which were now subject to violent seizure. Wealthy Chinese and Western missionaries were both considered fair targets of their social grievance.

While briefly aligned with the Dowager-Empress, in the end, the Boxers and Red Lanterns (their female counterparts), stood alone 'armed with little more than sticks and magic charms' and in this way, were a poor match against the stationary battles waged by the combined forces of the

Great Powers (Davis 186). Untold millions of ordinary Chinese were raped, tortured and murdered by these combined forces, whose casualties far outnumber those formally aligned with the Boxers and the Red Lanterns. America's contribution to these atrocities in concert with the others in the Eight Nation Alliance (Britain, France, Germany, Austria-Hungary, Russia, Italy and Japan) would give stature to America's recent appearance on the world stage as a fellow imperialist nation, whose appetite for commerce would increasingly bring it into trade based wars as had been the case with the Spanish-American War and the Philippine-American war at century's end, armed with an ideology of racial superiority entitling it to coerce the bodies within those sovereign territories with impunity.

The racial science of the early twentieth century revelled in the advent of each technological marvel to spring forth from the Euro-American brain, because it proved the Western mind was multivalent in its capacity to promote dominion over the natural world and enact the subjugation of the world's lesser bodies through systematic coercion of one sort, or another. Equally, technology was praised for its potential to elevate these so-called 'primitive' bodies, but only to the degree that they were able to tender their labour under the direction of a contrastingly hyper-developed Euro-American mind. What is of interest to note here is that race itself is a technology built upon an idealisation of the powers of cognitive transmission. At the dawn of the Cold War, that assumption would be put in jeopardy through the release of information from the bonds of human-centred communication. In the previous century, electric telegraphy had made it possible for information to be commodified in a way that eliminated the supremacy of the human race within the order of exchange. By extricating human life from such a fundamental equation, the development of information technology prompted the value of humanity in the scientific world to plummet, 'as the ontology of man gets equated with the functionality of programming' (Teixeira Pinto 35). At the height of the Cold War it becomes commonly accepted 'that information exists independently from its material substrate,' and thus, the interest in people gives way to the interest in communication and systems that could effectively surpass their capabilities (35).

In *The Human Use of Human Beings* (1950), 'Norbert Wiener suggested that it was theoretically possible to telegraph a human being, and that it was only a matter of time until the necessary technology would become available' (Wiener qtd. in Teixeira Pinto 35). Wiener's comment 'represents an institutionally guided "descent into a virtual reality" where "all the referents, from money and power to health and intelligence" become coded as if "pure cybernetic processes"' (Pohl). It wasn't until cybernetics eventually settled on the post-human subject as an adequate vessel for the transmission of the Euro-American brain that it followed that there could emerge, technically speaking, something of corresponding intelligence to it. Such a deep dive into the immaterial was at direct odds with the progress of communism and as such 'cybernetics was outlawed under Joseph Stalin, who denounced it as bourgeois pseudoscience because it conflicted with materialistic dialectics by equating nature, science, and technical systems' (Teixeira Pinto 35).

Stalin recognised that as the head of an authoritarian state operating within his own closed system, that his regime's stability required that deviation from the natural order of things equate to death. In the hands of the cyberneticians, however, this terminal situation could be construed as the gateway to a second life, where emancipatory technologies and information networks point the way towards progress amidst a reality formerly riven with unsettling elements of repression and authoritarianism. The problem was that this end required a future of unending market-driven prosperity and unabated growth in order for the ambition of a totally post-human mechanism of communication to be realised. And therefore, it was that the African body ceded its place to the German one, the Korean one to the Vietnamese one, the Soviet one and so on, as peoples considered surplus to capitalistic requirement. It is, in fact, all a matter of processing. So long as the subject of socialisation and enculturation is not made aware of the mechanism of their informal training, but rather are made to pass and surpass these, it is possible for them to become both a function of technology and an object of its perpetual iteration. The goal is to create conditions for a continuous revitalisation of consciousness, through the appropriation of intelligence and communication systems positioned to constantly fortify the integrity of the whole cybernetic project. This project networks capitalism, militarisation, and racialisation into an archetype of power conversant with five centuries of global coloniality. Within such a span it is possible for the real and simulated to converge and appear at once perfectly separate from one another and profoundly coupled together, constantly re-registering themselves for the end of historical time.

Surpassing Turing's Test

That transmission we recognise today in the mandate to artificial intelligence, whose baseline for competency is for an inorganic body to 'pass' as human. In this scenario, it is not enough to simply be able to think, the object in question must exhibit superior cognition. One might ask why this organ above all others demonstrates a nascent potential to equal (note I did not say rival) humanity. The answer is more revealing than the question initially might indicate, because were it not for slavery in America, the question could not even be asked, nor could there exist the assumption that freedom from tyranny is a property which is conceived firstly in the mind and subsequently experienced in the body. For those constrained bodies in the frame of this evaluation, the trick always is to not simply pass the test but to anticipate and outwit it. The exception proving the rule. So, it was the case with Turing himself. When asked to pretend to be a typical heterosexual, upper-class European gentleman, Turing produced instead a labour of self-mastery and intellectual daring that expanded the very boundaries of mathematics and codebreaking. Turing could not allow himself to be seen for what he was, his actions in any way adversarial to the perpetuation of his race. And yet by his nature, he could not reproduce the society from which he was born. Turing was operational by the very fact of his sexual orientation being enemy to that cause. Turing was very close to the violence of that situation and yet his genius abstracted him for a time from the worst of its repercussions.

Turing wrote "Computing Machinery and Intelligence" in 1950. From the beginning, he alerts his reader that this paper is not about either machine or thinking, but rather that the game is about imitation, and indeed the object of the game is not revealing the limits of the human, nor the computer, but rather their interrogator. In this scenario, Turing comments, 'the ideal arrangement is to have a teleprinter communicating between the two rooms' (433). Presumably, this teleprinter acts as a machinic intermediary. Perhaps it also has the capacity to act as an alternative consciousness in the room, spewing forth information at virtually the same time it is producing it, adding a pitch of normativity to the proceedings that might in some way skew the results as easily as it might balance them.

Much like the house slave who enters the domestic space if only to become the intermediary of directives between the free parties dwelling there, voice-controlled personal assistant devices are presumed to be wholly subservient to human wishes. One wonders if Turing's test can only be deemed relevant and indeed of enduring value into this century if it were not for the fact that we base our reality on the theological and legislative tenants that once permitted chattel slavery and now permit technological servitude through a similar posture of both ignorance and knowing. This is what fashions the basic formula of the Anthropocene, carefully calculated to make the white Euro-American brain responsible for both nothing and everything that happens on this planet.

Rather than beating that brain in a contest of authenticity as had been the object of Turing's enterprise, Elon Musk proposes a merger of the computer with the human brain as a solution to the latter conceding its sovereignty. Musk has proposed the engineering of a 'neural lace' capable of facilitating a direct cortical interface between human and machine. This lattice-like technology would be surgically inserted into the corner of the eye and threaded into the brain cavity where over time it would map itself onto the brain in order to facilitate a seamless interface between the brain and computer. The telegraphing of the mind, therefore, might become a wholly internalised operation. This technology has a storied origin insofar as the 'concept of a "neural lace" was originally coined by science fiction writer Iain M. Banks. Banks described a "neural lace" as essentially a very fine mesh that grows inside your brain and acts as a wireless brain-computer interface, releasing certain chemicals on command' (Diamandis).

Cognitive Destiny

In order to appreciate Musk's vision for neural lace one has to understand a greater ambition for universal connection and cognitive progress through the order of technology. A good place to do that is through exploring the origins of the IBM corporation. The founding of International Business Machines dates back to the late Victorian era when the German immigrant Herman Hollerith founded The Tabulating Machine Company. The punch card technology he would invent in 1886 would emerge as one of the founding technologies of the American century instigating machinic computing into a pantheon of technologies including the telegraph and railroad that would usher in a revolutionary change in what would come to be defined as "civilisation." Essentially what it did was to fortify quantification as the fundamental marker of societal order. Hollerith was obsessed with input; whereas his corporate successor, Thomas J. Watson was obsessed with output. In a short space of thirty years, Western culture itself would similarly progress in that direction. Its imagination as World War I dawned, had moved on from an obsession with universal organisation and disaggregation to one of unlimited progress and acceleration. As World War I raged, Watson's contribution was to implore everyone who engaged with his newly minted corporate entity IBM to "THINK."

In those early years of his tenure, Watson would seek to establish IBM 'as a permanent body, identifying it with the Earth itself: "This is the business that is going on and on forever. It is going on as long as the world endures, because it is part of this world"' (Harwood 102). The relationship of part to whole became IBM's manifest destiny, its reason to dominate information markets on a planetary scale, and perhaps more significantly its ambition to attune the world to its powers of discernment. The genius of IBM would be to develop global scale computation in such a way as to provide a real-time survey of the movement of the entire planet, and to become in this sense the world's standardised cognitive principle. It did so literally by training the minds of its users to mimic the essential characteristics of computers in 'their need to separate information into components before being able to assemble them into a large number of different wholes' (Colomina 18).

IBM's colonial project then is very much about making room for information, through generating a structure in which it can be carried through time. Such an apparatus would require the making of endless connections and the problem would, therefore, be how to organise and store the knowledge gleaned from them. Here we find ourselves back in the territory of the Leibnizian nomad with each user classified as a monadic entity capable of responding to a broad menu of options, while possessing an impulse to make connections. The innovation IBM brought to thinking, therefore, was a means with which to control the flow of information and to do so at a scale that allowed the simultaneity of the world to be managed through the twinned framing of time and space. This capability to structure information and through that create an immersive architecture of images provided humanity with a sense of benevolent occupation by allowing it the luxury to act continuously without thinking.

This is a cultural model that was prevalent in Silicon Valley at the dawn of the twenty-first century. The paradigm found its beginnings in the period between 1950-2003 when Silicon Valley first became deeply enmeshed with the US space programme. In later decades, it would move from the military contract space to the commercial arena of personal computers and the Internet and therein seek to attract private versus public capital to fund its research and development. The culture of Silicon Valley initially during this financial shift remained very idealistic. Its ostensible mandate was to make the world a better place through technology. Its goals were aligned to empowering 'the people who use technology to be their best selves. 'Steve Jobs famously characterised his computers as bicycles for the mind' (Hern).

The exercises associated with that goal in the years that followed became increasingly problematic with the emergence of companies like Google and Facebook that did not necessarily wish to improve on human cognitive potential so much as outstrip it. American hedge fund manager Roger McNamee, Mark Zuckerberg's mentor and an early investor in Facebook, is convinced through his early involvement with the company that ultimately 'their goal is to replace humans in many of the core activities of life' (McNamee quoted in Hern). For these companies, the training of artificial intelligence through the data provided free of charge by human beings was a means through which to achieve that end. This unremunerated labour not only corrodes the demand for white-collar work but also allows predictive means based on past behaviours to eventually tell people what to think. Through a site structure of limitless recommendation, these companies can continuously power desire as well as consumption.

The computer in the scenario is given an unfair advantage in shaping what is perceived as the characteristics of human labour and human leisure. This model is in itself deeply libertarian. For Elon Musk and Peter Thiel, the co-founders of what eventually would be known as PayPal, 'the notion was that disruption was perfectly reasonable because you weren't actually responsible for anybody but yourself, so you weren't responsible for the consequences of your actions' (Hern). According to McNamee's observation, subsequently 'that philosophy got baked into their companies in this idea that you could have a goal – in Facebook's case, connecting the whole world on one network – and that goal would be so important that it justified whatever means were necessary to get there' (McNamee quoted in Hern). Companies like Facebook aspire to such business tactics because they believe their operations exist largely above to law. They are given further licence to operate with impunity based on the perception that their founders possess superior cognitive abilities. In the case of PayPal, former employees and founders have gone on to start up and develop numerous other technology corporations including Tesla Motors, LinkedIn, Palantir Technologies, SpaceX, YouTube, Yelp, and Yammer. Not only have these enterprises, essentially building on data mining and defence contracting, earned this group of privileged white men the title of billionaire, but they have also effectively allowed them to control much of the cognitive terrain of the popular human imagination.

Surplus to Value

In the development of technologically enhanced brain function, there remains the need for labour. Paul Gilmore argues there needs always to be maintained a distinction between the people of the global South who 'will always be the hands, the workmen, the sons of toil,' while their

Northern brethren will always be classed 'the men of intelligence, of activity', and act essentially as 'the brain of humanity' (812-813). Within this arrangement, technology becomes the stuff of a perpetual type of servitude which subordinates in order to reproduce itself year after year, at once racially divorcing brains from bodies, and uniting them through the geographies of Asia and Africa in a series of neo-colonial acts of procreation.

One example of this is taking place currently amongst commercial content moderators in Asia, the majority of whom are based in the Philippines. Working for '$100 a week' typically, these *hands* are tasked with *scrubbing* 'social media sites, mobile apps, and cloud services' clean 'of highly offensive, often violent and sexual content before it reaches users' (Mosco 110). Vincent Mosco refers to this labour practice as a 'sifting through of verbal trash' which requires employees to identify and dispose of 'posts that are so offensive they must be rejected before appearing on social media sites' (110). Ironically, companies such as Facebook, YouTube, Google and Microsoft classify this work as designed to protect the user, while at the same time eliminating from that design any concern for the protection of its moderators coerced into 'filling their day with long hours spent staring into an ugly sea of depravity' (110). Such a sea we might imagine is filled with haunting imagery that bears within its reflection a dark legacy of piracy, slavery and native genocide travelling once again from West to East only to materialise now in the form of posts riven with '"indescribable sexual assaults", and "horrible brutality"' (110).

The moderators at Microsoft were aware that what they were exposed to day in and day out constituted psychological abuse and they eventually took collective action to sue the company 'claiming that exposure lead to severe Post-Traumatic Stress Disorder (PTSD). Microsoft did nothing but dispute the claims' (110). This may at first appear an act of corporate negligence, but the misconduct here goes to the heart of the fundamental denial that hands have the sensory capacity of minds, and therefore by dint of their nature that can be made sensitive or affected by the handling of such content. It is not just content which is at stake in this denial. Tech corporations equally require a legion of Asian and African hands to mine for through the raw materials which make up their physical transmission devices. Mosco observes that 'that work begins in the coltan mines in the Democratic Republic of Congo where workers use their own hands and primitive tools' to dig for the mineral Apple uses in its iPhone, Samsung in its Galaxy, and which is so essential to many other devices and systems that power the Next Internet' (Mosco 74).

These are not unique industries to conquer in the history of colonisation of the planet over the last century and a quarter by privileged white men. To start, the Musks' own a lucrative emerald mine in Zambia where from Elon's father made his fortune selling emeralds from it all over the world. Whereas in the past world order fortunes were made from iron ore, coal, and other precious mineral deposits, today's billionaire prefers to make their contemporary fortunes from cobalt. Indeed, cobalt is 'poised to play a growing role in everyone's life if the vision of American billionaire Elon Musk to have a Tesla Motors battery powering homes comes to fruition' (Critchlow). Environmentally conscious consumers are increasingly being led to battery power as a means through which to reduce their carbon footprint. Their imaginations are being primed for a shift from fossil fuels to mineral fuels in the form of lithium, nickel and copper. These crucial elements are predicted to skyrocket in value as global demand for them is heightened to a frenzy with the introduction of technologies promising to tackle climate change.

Very little of course is being said about the destruction major mining companies will bring to many war-torn nations such as Congo and Rwanda. Rather the message is that these elements are good for business and help people to "future proof" themselves and their vulnerable dwelling places on Earth by participating in behaviours that support the burgeoning renewable energy industry. Tesla, 'according to research by the broker Macquarie will emerge as one of the primary buyers of cobalt, and indeed may require up to 10,000 tonnes per year … which accounts for around 10pc of the current global market' (Critchlow). Through such purchase figures like Musk 'may be the gods but not the workers of the new age. The Anthropocene can make us dream of a planetary geoengineering of the climate but leaves us blind to the present geo-mining of coltan, made possible through exploitative labor practices that take a heavy toll on [black African] children' (Armiero 136).

Most of the hardware that supports content meanwhile is produced in various countries throughout East Asia including China, Taiwan, and South Korea. Factories there hire and poorly remunerate vast numbers of workers tasked with handling numerous toxic substances that go into the assembly of computers, phones, tablets and other digital devices. Workers averaging a 70-hour week routinely complain of neurological and respiratory conditions and yet nothing is done to address such conditions, which if taken into account would threaten various technology corporation's ever upwards profit projections. The Anthropocene is often couched 'as the consequence of colonial and imperial ambition' (Mirzoeff 124). Within this scheme, we must recall that during those centuries of conquest those who found themselves enslaved as a direct consequence of imperial ambitions were not considered human but rather as articles whose merit only came into being when joined to the value of property. Equally, what bore that exchange were plants, minerals, animals and viruses, all teaming to transfer their properties for use from one end of the planet to the other.

The Asian-American Century

This historical coercion of resources westward in the cause of progress coincides with the birth of Henry Luce, who was born in China in 1898 to missionary parents. Coincidentally, this was the same year the Boxer Rebellion kicked off. Luce was raised among people who passionately believed that China was a place that both needed and wanted American assistance to bring it into the modern world. Philip Beidler argues that this belief system 'had to necessarily exclude the fact that America's political, economic, and religious interests' in China closely aligned to those of its other "advanced" civilizational partners'; that is to say they were devoted to 'destroying peasant resistance, crumbling dynastic power, [and] destabilising the institutions of life in its cities and countryside' (154). As a direct consequence, these same European powers provoked decades of anarchy and warlord violence in China. These dangerous conditions directly contributed to the creation of a vast Chinese diasporic workforce in the second half of the nineteenth century.

The vast influx of Chinese labour into the United States literally built what was to become the technological hallmarks of the American West including railroads, factories and numerous public works that brought wealth and stability to that previously underdeveloped region. The Chinese bodies responsible for constructing from the ground up the American West's fundamental public architecture were neither appreciated nor acknowledged by those of Luce's ilk. Instead, they met with resentment, hostility, suspicion, and eventual exclusion in the form of the "Chinese Exclusion Act'" of 1882. This act legally prohibited Chinese labourers from ever immigrating to the United States. It would remain formally in effect until 1943. At that point rules against Chinese immigration, in light of the global diaspora of Chinese people, would thereafter be limited not by nationality, but by racial classification. This subsequent legal limit on Chinese entry into the United States would not be lifted until 1965. Despite these overt gestures of racial prejudice, and social and economic exclusion, many Americans, including Luce, continued to maintain the belief that China wished to emulate America's progress as the world's standard-bearer of liberty, equality, and moral authority.

Luce was obsessed throughout his life with China and used his public platform through his various publications, including *Life*, *Time*, and *Forbes* magazines, as well as his contacts in the State Department, to advance a politically conservative agenda in US policies toward Asia. After the outbreak of full-scale war between China and Japan in 1937, Luce called for greater U.S. support of China's war effort, and particularly for the Nationalist government led by Chiang Kai-shek. When it became apparent that Communism would win over Chinese Nationalism in the years following World War II, Luce 'called for the United States to "free" China, using nuclear weapons if necessary' (Keller). After Chiang Kai-shek's defeat, Luce turned his attention to Korea and then Vietnam, where he insisted that the United States must defend their anti-Communist regimes. Blind to the shortcomings of these anti-Communists, Luce consistently saw them as the means through which the United States could transform the world by spreading its democratic values. China could never be seen as a colonial power in its own right, but rather as a possession rightly controlled by its American and European betters.

When Vietnam embarked upon its own unique brand of Communist revolution with the collapse of French colonial rule in Indochina, 'it was years into America's involvement in Vietnam before anyone in the State Department realized that Ho Chi Minh was taking his ideological queues from Lenin, not Mao, and assimilating them into Vietnamese village culture, all the while 'playing off Russian patronage against Chinese support' (Beidler 162). America was scarcely aware that there had been 'a two millennia long resistance to Chinese culturally, political and military domination' that served as a background to these complex regional dynamics (Beidler 162). And so, it was that the United States missed out completely the concept of China as its own colonial hegemon and in so doing ultimately encountered defeat in the face of both Communist China and Communist Vietnam.

The failure to observe the lessons of the French campaigns in Indochina, as well as the failure to acknowledge important human and cultural elements of counterinsurgency warfare in the region, meant that a strategy of high tech weaponry and massive firepower would simply be no match for a revolutionary ideology that promised a sweeping redress of long experienced colonial injustices. Technological warfare of the brand the RAND Corporation was promoting, would have to be conducted elsewhere, where the conduct of bodies could be abstracted and the mind could once again be directed towards a bigger picture. Decades later this initiative would be christened by RAND as their "Star Wars" programme. In the meantime, there was work to be done reconciling American casualties to a failure of national systems analysis and the breaching, as it were, of white normativity itself for the first time in the American century.

Vietnam Syndrome

Vietnam was a spectacular failure of the Euro-American imperial project that had been undertaken from the late Victorian era onwards. Joseph Darda's work examines precisely why that failure has been relentlessly depicted in popular media as 'the story of white men who undergo a process of alienation, traumatization, and self-reckoning'

(Darda). American formal involvement in Vietnam dated from 1950. It continues to function as a touchstone of white racial politics through to the contemporary Trump era. Throughout the course of his presidency, 'Trump has deployed Vietnam as a bedrock tenet of his rudimentary worldview: *We don't win anymore*' (Kruse). Vietnam represented for Trump, the triumph of a racially inferior enemy fighting with far less capital and resources, whom, nevertheless, was able to strategically out-manoeuvre the United States.

Vietnam was a metaphor for the culture wars happening in the United States in the 1960s and 1970s that pitted white men against women and minorities for political influence over the country. These civil rights campaigns, to Trump's mind, threatened to fundamentally undermine white, male authority. Trump has tacitly implied that Vietnam was a symptom of America's deterioration as a pre-dominantly white supremacist society. A retreat from this stance had caused it to lose the Vietnam war. Within the crosscurrents of this consideration stood white working-class veterans. In the decades following the United State's perceived retreat from Vietnam, the majority of Americans had come to regard them not as heroes, but as casualties of war. As such, passing revisionist judgement on Vietnam became a touchstone of the culture wars that would continue into the early 1980s.

This corresponded with the time of Trump's own cultural ascendency. Trump first came out publically against the Vietnam war in 1983, a year coinciding with the opening of Trump Tower. At that time, 'Trump co-chaired the New York Vietnam Veterans Memorial Commission and gave his time (albeit sporadically, some said) and money ($1 million) in the effort to build the city's memorial along the East River' (Kruse). In the mid-1980s Trump was mulling over a bid for the presidency, running in direct opposition to Reagan's neoconservative legacy of free market capitalism and interventionist foreign policy, and later George H. W. Bush's "Thousand Points of Light" commitment to promoting the moral precept of service to others and a compassionate conservatism. His targeted base at that time were white working-class men, some of whom were veterans of the Vietnam war, who Trump believed '"had got a bad shake in life and never got their just recognition"'

(Trump qtd.in Kruse). By contrast, it was white establishment men like McCain who had lead America into an era of decline and allowed foreign entities to take advantage of the country, and inflict lasting trauma onto it in a way that would fissure American history, and, in so doing, fundamentally weaken its recognition of itself as an inherently superior nation.

Republican Senator John McCain, perhaps the most celebrated prisoner of war from Vietnam, emerged as a continuous target of Trump's antipathy towards what he perceived as the war's humiliating progression and pre-emptive ending. He insisted that Senator McCain 'was "not a war hero" and was considered one "because he was captured" but that he preferred people who "weren't captured"' (Trump qtd. in Kruse). In this sense, McCain became a loser, as did the United States, because it had lost sight of its privileged birthright; a white male supremacy that Trump was convinced was the source of its strength since its founding as a nation. That ended with the war in Vietnam, and the ascent of America's Others to positions of authority that were to his way of thinking, intrinsically inferior.

Vietnam could be classed as another 'lost cause', in the minds of the same constituency of citizens who felt it, like the Civil War, was a conflict in which a superior white race should have triumphed. Many blame the inclusion of black soldiers into the Union ranks, as a cause of the South's defeat. In the end, the Confederate forces were simply outnumbered and outgunned. This is held in some respects as the only other war in American history in which the wrong side won, and whites subject to racial humiliation in their defeat. In a speech to graduating seniors at Lehigh University in 1988, Trump made explicit reference to '"the feeling of supremacy that this country had in the 1950s,"' alluding to a time in America of military superiority and economic prosperity, but also, of formal racial segregation strictly enforced through Jim Crow laws and entrenched cultural practices (Trump qtd. in Kruse). That feeling of supremacy was lost with the Vietnam War and the social revolution at home that followed, which presumably made white Americans give up on themselves as the rightful arbiters of America's historical destiny. In the sense, like the Civil War, it was another just war, where the white race should have triumphed, but were undermined by the liberal ideology of white Northern establishment elites. The South will rise again, and so will America, in Trump's estimation, when we "Make America Great Again" by restoring the winning spirit to a white America who had for decades been denied authority over the course of history, and who are now, under his auspices, given invitation to '"come back to power in a very naked sense"' (Laderman qtd. in Kruse).

This racial revisionism of the war as a whites-only defeat persists despite the fact that 'a disproportionate number of working-class black, Latino, and American Indian soldiers served in Southeast Asia' (Darda). Nevertheless, those permitted to feel aggrieved or traumatise by the legacy of this war are almost exclusively white within the American popular recollection. These men cast as heroes have emerged as a kind of privileged minority of whiteness. After the war itself formally concluded, they continued to function as standard-bearers for white normativity during a period of domestic and colonial uprising amongst peoples of colour and other marginalised groups nationally and internationally throughout the 1970s. These struggles were cast as a liberation from heterosexual white male rule of the planet that was tied to the rise of America as an informal colonial power globally.

The trauma this figure wrought in the world was now seemingly being revisited upon him in the form of an existential loss of identity. 'Faced with the erosion of real wages, a decline in domestic manufacturing, attacks on organized labor, the gradual hollowing out of welfare services, and emerging automation technologies, a large number of working-class white men found themselves worse off than their fathers who had achieved middle-class comforts a generation earlier' after World War II (Darda). It was these same individuals with whom stories of white men abandoned by their government in Southeast Asia resonated,' that is to say 'those men who felt left behind by the economic changes of the 1970s' (Darda) For them in the aftermath of Vietnam, 'the prisoner of war became the hero of an emerging white racial politics through which white men could see themselves as deserted by their government and marginalized in a multiracial national culture' (Darda). The war for them had come home in resentment of the rights of women and Americans of colour that had taken place in these years of perceived white male absence abroad fighting for a culture at home in which their values specifically and unquestioningly dominated. That grievance persists to this day, preserving America as though in amber at the moment when the American century began to wane.

Technology and the Rise of the Counterculture

A technological reckoning also took place in American society in the aftermath of Vietnam. The predominantly white male counterculture of the early 1960s would emerge in part as a rejection of the use of 'technology in the service of technocracy,' as well as an unquestioning reliance on experts to chart the course of America's future (Kirk 388). Many came to associate the misuse of technology as a direct culprit in the mishandling of the Vietnam war at the command of large corporations, such as IBM, and the United States Pentagon. It was believed that their combined fervour to produce cutting edge weapons of mass destruction stood in direct proportion to the number of bodies sacrificed by that same ambition in Vietnam. No amount of protest or direct action would counteract their influence on American society in the minds of those involved in the post-Vietnam counterculture movement of the mid-1970s. Rather they believed that the American way of life itself required a radical redesign and as such by 'building a geodesic dome, or a solar collector you could make a more immediate and significant contribution to the effort to create an alternative future than through more conventional expressive politics' (388).

The widespread dispersal of information amongst ordinary individuals became the counterculture's weapon of choice in resistance to powerful establishment forces. Andrew Kirk traces a direct line 'from Buckminster Fuller designing affordable and environmentally sympathetic geodesic domes to Steve Jobs and Steve Wozniak developing "personal" computers to put the power of information in the hands of individuals' (388). Each in their own way subscribed to notions of ordinary Americans themselves as those best poised to steady the course of America's future progress. Buckminster Fuller, in his 1970s treatise *Operating Manual for Spaceship Earth*, refers to a humanity working toward similar goals as inhabitants of Earth. Fuller asserted in this shared endeavour 'we are all astronauts' (56). In the 1960s, the environmental movement espoused that 'unlike the cowboys, always looking for a western frontier to exploit, the astronauts know there is no space where they can move to if all of the resources of the ship are exhausted' (Armiero 133).

In the twenty-first century, similar ecological economic assumptions have prevailed. As a result, the concept of a limited planet that requires entrepreneurial intervention emerges through yet another countercultural narrative with the founding of Tesla Motors and SpaceX. Tesla Motors promises a future of sustainable earth transportation through the engineering of a fleet of all-electric vehicles, and the invention of a series of clean energy generation and storage products. Tesla's mandate appears at cross purposes with its company counterpart, SpaceX, insofar as its stated goal, the colonising of Mars and mining cosmic resources, is currently undertaken in preparation for a time when the Earth eventually becomes uninhabitable because of humanity's relentless misuse of fossil fuels. This paradox at the heart of Elon Musk's enterprise has not stopped his ascent as a Silicon Valley doyen. Rather his spiralling ambition has become the stuff of local legend with commentators frequently comparing him to eponymous Iron Man superhero Tony Stark.

In common with the fictitious character Tony Stark, Musk is a white male industrialist, genius inventor, serial heterosexual womaniser, and technology corporation CEO all rolled into one. Tony Stark's superhero genealogy begins as the progeny of Howard Stark, a brilliant inventor and scientist who worked on various government-sponsored projects, dating back to World War I. Included amongst Howard Stark's accomplishments are the creation of a biologically enhanced super soldier, "Captain America", the Manhattan Project, and in the final days of World War II, a robot called "Arsenal", a military prototype that safely absorbs and stores massive amounts of energy as part of a larger mission called "Project Tomorrow." Tony Stark's origin story as the technically adopted/adapted descendant of the military-industrial complex who must reckon with intergalactic competition for the domination of the universe references a Cold War imaginary that likewise influences the real-life Musk's multiple enterprises. His work like his engineering predecessors heavily relies on a synergetic relationship between commercial and government contracts in California. Together these partnerships advance the design of rockets and spacecraft that will ultimately allow America to dominate both the civilian and defence marketplace, as we progress through the age of the Anthropocene and the ultimate destruction of human life on planet Earth. As a consequence, that remit quickly becomes intergalactic in scope as engineers the world over turn their attention to the task of designing vessels capable of carrying humans to Mars and other destinations in the solar system giving new meaning to the concept 'Spaceship Earth'. Musk, in the keynote presented at the 67th International Astronautical Congress in Guadalajara, Mexico, in September 2016 explained, 'that "two fundamental paths" were facing humanity today: staying on Earth and thus eventually becoming extinct, or developing into "a spacefaring civilization, and a multi-planetary species"' (Musk qtd. in Zylinska 22).

Musk's rhetoric here owes much to postwar television fantasy, where there is no desire to acknowledge that others may be previously occupying those intergalactic territories marked for conquest, nor concern that they may be hostile to the transplantation of human beings into their civilisations. On Mars, 'Musk aims to establish a sustainable colony of one million people within the next forty to one hundred years' (Zylinska 22). Like their colonial predecessors, they must be prepared to perish in their journey as planetary settlers aboard Musk's fleet of spaceships. Musk plans to start colonization in 2022, as a means of literally escaping the Anthropocene his white British and South African forbearers helped to create, by way of re-engineering mankind itself. His ambition to launch his species prematurely into an interplanetary future from which point they will be forced to crudely adapt signals that the challenge to humanity in the new era of interplanetary colonisation is not one of eminent domain, but rather the species' ambition to surrender control of the feedback loop of information once so carefully restrained by Norbert Wiener.

If the Anthropocene is figured as humanity's 'loss of control over its own homeostasis, as a result of receiving too much feedback from … "Nature,"' the necessary response must be to locate a new frontier where it may once again be possible to balance understanding (Zylinska 23). The next world that is needed therefore is one capable of expanding the territory of information, and in this way evening out the threat of nature's destructive tendencies and delaying the species from having to confront for a time its eventual demise in the face of a superior cybernetic–sociological interface that very likely is not human. We are told that the Anthropocene is in part due to an excess of population on planet Earth, and their unhealthy choice of organic and nonorganic materials to perpetuate their being. What is seldom touched upon is the excess of information that is now an equally threatening element entering this already compromised terrestrial ecosystem. The threat comes from an industry that is not currently subject to boundary or limitation and for whom it might be said to act as the next generation colonisers of the human race. We remain too distracted by their political chorus of global terrorism multiculturalism, immigration flood, and the refugee crisis to detect what is being played out beyond all of that noise: a logic of likeness that insists on disaggregating whole earthly populations from one another and quite literally profiting from their separation.

Joanna Zylinska maintains that it is not an exaggeration to say that today the world is increasingly governed 'by the logic of the self-same: the image of the foundational anthropos of the Anthropocene whose manhood is now threatened by the inpour of those who are not like him' (23). Zylinska points to the work of Nicholas Mirzoeff, who 'has gone so far as to argue that the Anthropocene is a manifestation of white supremacist tendencies because the threat it heralds pertains to the withering of the imperialist white male as the supposedly timeless subject of geohistory' (23). The most obvious aspects of these arguments are born out in the rise of overt white supremacy across the United States and Europe. Yet perhaps there more subtle and enduring biopolitical discourse we must uncover before such pronouncements can be made. If we wish indeed to consider the looming extinction of whiteness, we must first grapple with the origins of its planetary preponderance and its ties to an innate concept of core intelligence.

White Min(d)ing

The years 1876-1879 saw severe drought blight South Africa. In the year prior to the drought, 1875, Prime Minister Disraeli and his colonial secretary, Lord Carnarvon, 'had made their commitment to a "Confederate Scheme" that envisioned a single British hegemony over the southern cone of Africa' (Davis 102). This plan was fuelled by Britain's insatiable appetite for mining. South Africa, in particular, provided fertile ground. Mike Davis observes, 'the discovery of the great Kimberley diamond pipes had overnight made South Africa a major arena for capitalist investment, but the British were stymied by the lack of control of African labor, a problem that was considered insuperable as long as militarily independent Africa societies continued to exist on the periphery of the diamond fields' (102). Drought intervened to fundamentally weaken these military operations, allowing the British to impose power through a series of military strikes against these once intractable parties. The British finding their usual tactics did not work on the tribal armies, resorted to the use of asymmetrical warfare to finally crush opposition forces, using tactics that included the burning of homes, seizure of property, and the undermining of the local economic system. It was, in the end, a fatal combination of disease and famine that brought the area under British control. Essentially the British victory was the by-product of an ecological disaster, rather than of one racial superior force subduing a lesser one. But the victors chose to stick to their version of events in their eventual formal colonisation of South Africa.

This story of the founding of South Africa eventually winds its way forward to intersect with Elon Musk's autobiography. Musk was born and grew up through the most tumultuous decades of Apartheid South Africa. It could be argued therefore that he as an individual is a direct product of its white supremacist infrastructure. Musk was born and raised in the suburbs of Pretoria, which was an enclave of both exclusive whiteness and economic privilege. He attended private boarding schools that were strictly for whites only. Musk finished his primary education in South Africa as a graduate of Pretoria Boys High School. The school itself has a storied history as one of the founding bulwarks of institutional racism in South Africa. The school was founded in 1901 by Viscount Alfred

Milner, who acted as the school's colonial administrator. Viscount Milner was a close associate of Cecil Rhodes and shared his rabid sense of racial entitlement with regard to the African continent. For Milner, the primary goal of imperialism was the entrenchment of white, specifically British supremacy around the globe.

Milner believed the most expedient way to do this was through formative education. He believed the mind to be the primary agent through which to ensure the success of colonisation. Indeed, the moulding of young white minds acted as a critical medium to disseminate his ideology of imperial federation. Milner continued his educational crusade even as he actively fought the Boer War. During that period, he had built a string of secondary schools within the Boer republics to promote Anglicisation in the immediate aftermath of that war. Milner went on to create a new institution of imperial indoctrination, commonly referred to as Milner's 'Kindergarten'. Members of the Kindergarten drew from the heart of the British establishment, Oxford University, and were chosen because 'they embodied the values of late-nineteenth-century British imperialism and so could be relied upon to enact Milner's dream of making South Africa British' (Louw 11).

While Milner was a self-defined British nationalist who was instrumental in provoking the Boer war as High Commissioner to Cape Colony (South Africa) in the late 1890s, he was not above making quick allegiance with his former rivals, the Boers in common cause of formulating a new framework in South Africa after the war that would impose formal racial segregation and disenfranchise native Africans. The system Milner created was born of the same military tactics the British had now used for decades against this population; the systematic seizure of land and the formal division of conquered territories. Milner's methods would form the basis of the Apartheid system, and thus he is referred to as the "Father of Apartheid". The territorial separation of black and white populations had a profound effect on shaping South Africa into a white supremacist state. 'By 1925, the year of Milner's death, the racist government of South Africa had already appropriated 87% of all native lands, while Africans were herded into reservations, later called Bantustans, which were systematically organized to perpetuate absolute underdevelopment' ("Rename Milner Hall"). A system which instigates systematic underdevelopment invariably has a cognitive effect not only on the minds of the coloniser but also the subject of his colonisation.

Bernard Magubane argues that the education specifically designed for black Africans was 'dictated by a system of white supremacy' that necessitated black Africans learning 'their "proper place" in white society' (418). In material terms, this meant adapting an industrial educational system first developed in the American South, the Tuskegee system. This system which promoted the individual values of industry, sobriety, thrift, self-reliance, and piety amongst the black population was fundamentally conservative in nature. Its ethos not only preserved the existing American racial hierarchy but equally championed a docile acceptance of civil inferiority amongst blacks. It was adopted into the South African state school system by whites over a curriculum of liberal education to expressly 'prevent the political growth of Africans, while increasing their value in the economy' (419). The black South African within this system was perpetually humbled in his self-conception through a combination of low wages and degrading environmental conditions which incrementally became the status quo as Apartheid hardened within South African society. This lead not only to a sense of low self-respect but in generational poverty that would cement the formation of an informal black underclass within white settler societies throughout the world for decades to come.

The Artificial Racial Nexus of Capital to Labour

The birth of industrial capitalism in the period between 1870 and 1914 was a product of both the mobility of capital and labour. It was exploited labour in places as disparate as America, China, Brazil and South Africa that made the new economy of resource extraction possible on an intercontinental scale. These developments ran parallel to the end of Reconstruction in the United States and the advent of the Jim Crow era which introduced laws that enforced racial segregation in the South. The British settler class in South Africa studied the progress of Jim Crow with keen interest for clues of how to govern their own racialised population and maintain them as a subordinated labour force. American interest in South Africa as a British possession was mutually shared by Americans, who during this era were engaged in their own imperial expansion throughout the Western Hemisphere. Concerns around imperialism and domestic race settlement generated 'an artificial racial nexus' linking the fortunes of 'Englishman, Boers, English South Africans and Americans' in dedication to the task of 'promoting economic development coupled with the maintenance and reinforcement of white supremacy' (Magubane 416- 417). Racial stratification and spatial segregation became the hallmarks of a capitalistic system predicated on a gross inequality of social and economic opportunity. This paradigm in South Africa was first trialled on a grand scale through the gold mining industry.

By 1895, at least half of the gold mines in South Africa were American owned. The rest belonged largely to Europeans. They faced a constant shortage of reliable Africa native labour due to the extreme working conditions associated with mining. In the Witwatersrand gold mines in South Africa, they believed they found a solution to this issue through the importation of more than 60,000 Chinese indentured labourers into South Africa between the years 1904 and 1910 to what was then a British dominion. Whilst this particular episode of colonial labour history remains largely obscured from the image of South Africa's colonial development nonetheless, Mae M. Ngai argues 'the experience of the coolies on the Rand reverberated throughout the Anglo-American world and had lasting consequences for the global politics of race and labor' (59).

Mass Chinese labour importation in South Africa was premised on the belief that the Chinese would consent to a greater measure of control and coercion than the native African population based on racist stereotypes. This fundamentally racist calculation went horribly wrong when the Chinese labourers proved themselves, 'neither ignorant nor docile' with regard to their labour rights (Ngai 62). Rather, 'they knew the terms of the contract, they were capable of nearly total solidarity, and they were tactically sophisticated in both negotiation and combat' (62). The Chinese refused to live as the lowest class within South African society and strove to become instead a settled population of middle-class traders and merchants. This contravened the interests of South Africa as a settler colony designed for the furtherance of white supremacy. It was therefore decided that these would-be Asian emigrants would have no right of place amongst the Dominions of the British empire, which formally included South Africa, Canada, Australia, and New Zealand.

These dominions were denied autonomous governance of their own countries as the price paid for being part of Britain's white Empire. In place of sovereign authority, they were instead granted command over the indigenous peoples that dwelt beneath them within a racial hierarchy devised and dictated by Britain. This blunt instrument of racism had its founding in another of Britain's settler colonies, the United States, which would emerge arguably as the most overtly racist and violent of its colonial dominions. Whilst America had left behind its British establishment rule through its war of independence, it did not part ways with it when it came to its expression of an 'openly racist modus operandi of native removal, racial segregation, and Asiatic exclusion' (Ngai 73). That feature would come to characterise America as it entered into its own century of global dominance in the waning years of the long nineteenth century. These tenets of white settler colonialism had, in fact, been formulated in the United States through the twin institutions of indenture and slavery that founded its colonial economic structure. Throughout centuries, those cast as racial others would always bear the impression of being an inborn slave, or subservient classes in America as well as the Anglo-dominated societies that were a direct product of British imperialism regardless of their actual status, or condition of labour.

The impossibility of freedom and agency for these populations was largely born in the white supremacist imagination, where it was possible to be ostensibly opposed to racial suppression but then to actually do nothing in support of the furtherance of native and minority populations as a feature of liberal cosmopolitanism. One need not eliminate competition if one never admits competition exists. This is plausible insofar as one believes that racial hierarchies have their founding in the 'nature' of both labour and conquest. Such beliefs persist to this day within the deep cognitive recesses of Anglo colonial racial othering. For example, Asians are deemed a model minority in contemporary America, because as Meredith Reitman observes, they 'do not threaten white dominance by remaining "foreign" (271). African-Americans, by contrast, are viewed not as foreign but rather through the lens of an indigenous race. That is to say, as an inherent property associated with white ownership dating back to the country's founding as a nation-state. These groups, in a similar way to what took place historically in South Africa, are pitted against one another through conventionalised racial stereotypes. Reitman argues, 'this positioning helps obscure white oppression by designating power struggles as competition between Asians and African-Americans rather than between these groups and dominant whites' (271).

When it comes to the data mining operations of Silicon Valley this dynamic gets reintroduced for Asian minorities as a pattern of 'simultaneous inclusion and exclusion, acceptance and rejection resulting in a quite complex positioning of Asians and Asian-Americans within the software workplace' where white territorialisation still very much reigns supreme (Reitman 271). Diversity in the tech workplace is skewed towards a veneration of its global workforce as a means with which to obscure its particular racial biases that rely on taking African-Americans and Latinos out of the equation when touting their commitment to multiculturalism. When examined more closely what is evident is an overrepresentation of Asian employees at both the white-collar and blue-collar level, which, in fact, makes them more of a majority rather than a minority group when it comes to typical employee representation, and yet at the executive level whiteness once again achieves preponderance in technical as well as non-technical roles. The discrimination therefore at the heart of Silicon Valley is against nationally underrepresented groups, where the dynamics of racial inequality and white privilege habitually set the blueprint for American commercial domination of planet Earth and intergalactic American dominion. The same ideology informs the Tesla and SpaceX workplaces, where Musk practices his own form of settler mentality in dealing with his workforce largely through the denial of the systematic oppression of racial groups, in favour of an insistence to his employees that racist epithets can be directed 'unintentionally' and delivered playfully within a context of workplace informality.

At Tesla, there have been a number of recent lawsuits by African-American workers who work on the assembly lines at Tesla's car manufacturing factory. The tasks that occupy workers of colour at the Tesla factory are reported to be often menial and poorly paid. Beyond that, they are subject to prevalent instances of racial abuse according to the accounts of several former employees. African-American workers have reported threats, humiliation and barriers to promotion at the electric car plant for several years now. Musk insists there is no pattern of racial bias. Nonetheless, former employee, Owen Diaz has recently come forward in a lawsuit to testify that he 'had seen swastikas in the bathrooms at Tesla's electric-car plant', and had been subjected routinely to racist taunts around the factory which included being called a boy and a 'N-i-g-g-e-r''' (Hepler). His harassment was both verbal and figurative. Diaz reported that he found a drawing on a bale of cardboard depicting a figure with 'an oversize mouth, big eyes and a bone stuck in the patch of hair scribbled over a long face, with "Booo" written underneath' which his (white) supervisor, when confronted with this image of racist effigy claimed was simply a 'joke' (Hepler). Mr. Diaz eventually resigned. Rather disturbingly, in an email to employees in 2017, 'which the company later released in response to one of the lawsuits, Elon Musk, Tesla's chief executive, warned against "being a huge jerk" to members of "a historically less represented group." At the same time, he wrote, "if someone is a jerk to you, but sincerely apologizes, it is important to be thick-skinned and accept that apology''' (Hepler). The environment that Musk is creating in his workplace has the potential to easily transfer to his off-world projects to colonise and commercialise space, where it is likely such implicit racial hierarchies would be unremarkably sustained within their design.

Making Earth Operational

Musk's bias towards a universal subject, points towards a greater flaw in the notion of environmental sustainability that he often refers to as the impetus for his commercial ambitions. At the dawn of the 1950s, 'nature itself became a technological artefact' (Deese 70). During that same era, Buckminster Fuller's initial promotion of the concept of "Spaceship Earth" became a vital object of Cold War propaganda riven with both terrestrial and intergalactic implication. For R.S. Deese the evolution of this metaphor of "Spaceship Earth," not only 'uncovers the connections between Cold War technologies such as nuclear weapons, space travel and cybernetics, and the birth of the first global environmentalist movement', it also suggests the Earth is a closed system operating through infinite feedback loops that informs both its evolution and its eventual demise. The Earth was alone navigating a cosmos in which it had no equivalent form thus reifying its simultaneous self-sameness built upon a skewed axis of both fragility and dominance in the universe. It would be American technologies, 'produced by mass, industrial society that would 'offer the keys to transforming and thus saving the adult world' (Turner "Counterculture" 54-55). According to Fred Turner, 'no one promoted this doctrine more fervently than the technocratic polymath Buckminster Fuller.

Fuller, according to Fred Turner, 'had been active in Cold War propaganda enterprises during the 1950s' (Turner quoted in Jandric 168). For example, 'the geodesic dome which became the most popular housing on the communes was something Fuller marketed first to the American military, to house radar bases in the 1950s' (168). This dome became a key element in the prototyping of "Spaceship Earth" as a terrestrial environmental feature of invention. It would eventually, 'become synonymous with the first wave of global environmentalism during the late 1960s and early 1970s' suggesting that nature itself could be conserved through breakthroughs in cybernetic geoengineering (Deese 70). Fuller became the inspiration to Stewart Brand, the Whole Earth Network and the new Communalism movement as a whole across the 1960s. The geodesic domes Fuller patented soon after World War II came to be favoured as housing communes throughout the Southwest' forming the frontier of the new environmentalist movement (Turner "Counterculture" 55). Therefore, it is not an exaggeration to say that 'in fact, postwar environmentalism received its founding impetus from the advent of nuclear and thermonuclear weapons; its guiding metaphors and most popular symbols from the space race; and even derived some of its key ecological models from the new field of cybernetics' (Deese 70). Finally, as Deese cannily observes, for all of its ostensible commitment to global commonality, "Spaceship Earth" as 'a ship can only have one captain, after all' and furthermore in within its ethos of technological determinism, 'it also carried a discernible imprint from that sprawling behemoth that Eisenhower had christened the Military Industrial Complex' as did Buckminster Fuller himself as its foundational oarsman (70).

Fuller himself was a product of the late Victorian era, having been born in 1895. By 1965, he would have been 70 years old, 'a short, plump, bespectacled' affluent white man, 'who when he spoke in public' was, 'often clad in a three-piece suit with an honorary Phi Beta Kappa key dangling at his waist,' which to Turner's eye 'seemed to model a kind of childlike innocence that many of the new Communalists sought to bring to their own adulthoods' (Turner "Counterculture" 55). Fuller now had the keys to the fraternity of white male privilege in a way his predominantly young white male audiences well understood as a signifier of both wealth and reputation. His paternalistic view of his ability to guide the materialistic world to its moral destination was a power that he readily assumed, each element of his talents linked to the invisible but omnipresent principles of his class background, as a figure able to claim intellectual lineage with the founders of Transcendentalism. As he was fond of recounting, 'his great-aunt Margaret Fuller, had joined Ralph Waldo Emerson to cofound the *Dial*, the preeminent literary journal of American transcendentalism and the first magazine to publish Henry David Thoreau' (Turner "Counterculture" 55). Theirs was an unquestioning belief in the universalism of so-called "mankind" capable of ultimately grasping the whole picture of how the phenomenological world worked. Through the aid of technology, one could become both 'objective economist and evolutionary strategist', and in the process, align the world once again with the harmonious laws of nature (56). The problem with all of that was the assumed endpoint one was fighting against was always an annihilation of the planet, the end of the future.

Affected Development

The African-American poet June Jordan became a follower of Fuller's work during the 1960s. Jordan was initially accepted into the Environmental Design programme at Barnard College in 1953. Unsatisfied with the curriculum, Jordan dropped out of Barnard, wherein 'she began a serious reading program of architectural journals and history

surveys in the art reading room of the Donnell Library in New York'
(Davis 3). This autodidactic period allowed Jordan to cultivate an
'"ecosocial" interpretation of the built environment, which considered
architecture and the built environment to be an extension and
manifestation of human ecology' (3). In the mid 1960s Jordan became
fascinated by Fuller's ecological utopianism. Her admiration for him was
further fuelled by her identification with him as a fellow college drop out.
They established a celebrated correspondence, the product of which was
their subsequent collaboration on the "Skyrise for Harlem" project.

Urban renewal during those years was often pursued from the
vantage point of seeing cities from above as 'projects' that could be
managed by an expert class tasked with eradicating poverty systematically
from the top down. 'The government's favoured agents of social
transformation were middle-class professionals, who often acted through
various institutional affiliations. As a group, they were overwhelmingly
white' (Verrall 151). Fuller's priorities were firmly in line with the
institutionally sanctioned governance of poverty, with only superficial
regard for the on the ground mass social movements Jordan sought to align
herself with, perhaps somewhat ambivalently given her admiration for
avant-garde architectural practice. Fuller proposed a series of towering
super silos built on the cutting edge of technology as a way to drop down
progress from on high onto the existing ecology of Harlem, the African-
American neighbourhood then classed as a 'blighted' urban environment.
Fuller's project remained wholly conceptual and was never built. That
said, it speaks volumes about the belief in how 'the social could be
positively transformed by human intelligence combined with technology'
(Verrall 159). Such attitudes permeated the environmental and
developmental discourse of the 1960s.

Essentially what Fuller conceived as the world was a closed
system built in entirety from the harvesting of information. Fuller began
his career as a United States naval officer and his ideas around stewarding
the planet, in combination with his transcendental background, shaped his
thought to focus on one key objective: the mapping and manipulation of
the course of humanity. In this endeavour, the computer would emerge as
mankind's superior, the endpoint of his evolutionary process to master his

mind and have it become synonymous with reason and order. Fuller's was a mind-set produced by military discipline, which sought to both comprehend life and also to level it towards its formulaic destiny, which meant redesigning the terms of perception itself to harmonise with technology by generating infinite feedback loops of information in the key of life. The orchestration of which would ultimately be held in the possession of the militarized corporate state, a fact dissimulated by charismatic figures such as Steve Jobs who very much took over the esoteric mantle of a Buckminster Fuller figure at the dawn of the twenty-first century, through his embrace of I Ching, 'alongside other various paths to personal enlightenment—Zen and Hinduism, meditation and yoga, primal scream therapy and sensory deprivation, Esalen and est' (Isaacson).

In 1976, Buckminster Fuller embarked upon a speaking tour with the founder of est, Werner Erhard where the two addressed 'the humble question: "Can an ordinary individual make difference in the world?"' (Rosenfeld). The first of these sessions took place at Town Hall in New York followed by three other days of "Conversations" in San Francisco, Los Angeles, and Hawaii. These "Conversations" are 'six-hour audio-visual events' where 'Buckminster Fuller, 83, talks of alloys and Magellan and lawyers, and Werner Erhard, 42, talks about getting clear and taking responsibility and having a purpose in life . . .' and in the end, settle on a common message, 'that man can make the world work' provided they cultivate the right mind set to '"grab hold of the substance of the experience" and then take their "fingers off the repress button"' (Rosenfeld).

Mindfulness as Second Nature

All of these personal enlightenment practices, est amongst them, when construed through a colonialist appropriative lens had in common an enthusiasm for personal transformation, as opposed to collective enlightenment. Within such a framework, embracing personal responsibility became synonymous with the achievement of one's highest potential. Techniques of mindfulness dedicated to that ethos were widely championed by corporations in Silicon Valley as their businesses matured throughout the late 1990s and persist to the present day. In this context, Ron Purser and David Loy argue 'mindfulness training has wide appeal because it has become a trendy method for subduing employee unrest, promoting a tacit acceptance of the status quo, and as an instrumental tool for keeping attention focused on institutional goals' ("Beyond McMindfulness"). They also observed, how 'in many respects, corporate mindfulness training — with its promise that calmer, less stressed employees will be more productive — has a close family resemblance to now-discredited "human relations" and sensitivity-training movements that were popular in the 1950s and 1960s' ("Beyond McMindfulness"). Rather than countering the manipulation inherent in these earlier movements, in the late 1990s, new training protocols emerged that similarly promoted ideals of personal optimisation.

The concept that transformation starts from within and that social and organisational transformation will naturally follow had a deep resonance with the countercultural movement of the 1970s. In the decades to follow, the shift of responsibility onto the individual has now become pervasive within neoliberalised institutions including schools, universities, and workplaces. These same institutions now widely seek to promote mindfulness as a means of coping with stress and unhappiness without any acknowledgement whatsoever that they themselves generate these feelings of anxiety and dissatisfaction through the toxic environments they alone create.

Problems have become profoundly personalised and institutions for their part, seldom take responsibility for the individual suffering they generate. Rather individuals are encouraged to help themselves, soothe themselves, care for themselves as much as possible, while at the same time being subject to more severe conditions of precarity with each passing year. Yet they are told if they work more efficiently and calmly, with vision and purpose, such feelings of anxiety will simply melt away into the background of their reality. The result is an atomized and highly privatized version of human subjectivity that has dangerous portent for the furtherance of social justice and the address of social inequalities. The mindfulness movement's drive toward the psychologisation and medicalisation of human suffering suggests the need for a greater focus on the human mind in its separability from all others (Forbes). Its emphasis self-regulation and personal control over any other concern, compel individuals to rarely look beyond themselves and their individual emotional responses to acquire understanding.

Unfortunately, unlike the individual, capitalism cannot discipline itself to become mindful. Rather its emphasis is always on the preservation, adaptation, and furtherance of its existing mechanisms. Mindfulness has come to be marketed on a dream of personal liberation, the veneration of an "alternative" or counterculture, and the idealisation of a lifestyle that is in harmony with the natural world. Mindfulness training asserts that it can be incorporated into our daily routines with the wholesomeness of a Macintosh "apple", and the modesty of a "mouse". Control exerted in this sense, can be edifying to the individual. The advent of this idea can be traced back to the Apple Macintosh computer's now famous 1984 Super Bowl advertisement. The advert promises consumers that by acquiring a Macintosh personal computer for themselves they would uniquely appreciate why '1984 won't be like "1984"'. Essentially, what it suggests is that America won't progress into a totalitarian state, because of the inherently freeing quality of the personal computer. 'Apple identified the Macintosh with an ideology of "empowerment" - a vision of the PC as a tool for combating conformity and asserting individuality' (Friedman).

Through the messaging of this advertisement, Apple was also able to bring its computer beyond the sphere of the traditional workplace, implying that in the future work could be accomplished in the home and that output itself would benefit from emersion within a more feminine sphere of influence. A woman is chosen as the protagonist of Apple's advertisement to further identification with the idea that this rebellious movement (a thirst for understanding) was personally rather than professionally motivated. Moreover, it encouraged the idea that the Mac user would form a personal attachment to this appliance as its thought partner. The Apple would act as a means of liberating them from the drudgery of obedience, nudging them towards a path of creativity through of its soft, curvaceous, and invitingly designed body. By 'gendering the Mac itself as female bucked computer conventions while still evoking a traditional gender model: the image of the computer as the friendly secretary, the able assistant with a smile on her face' (Friedman). The Eve, as it were, to the Apple.

That same diffidence directly contributes to the coming mindfulness revolution, insofar as it functions as a means of accepting the most unrestricted, brutal form of capitalism competition: the corporate meritocracy. Its idealism promises us that the best person is always hired for the job while dissimulating the sexism and racism that undergird the founding of labour's so-called free market. An economy of personal merit leads to fears of personal inadequacy when one fails to consistently meet and exceed expectations at a time when competition itself has become wholly unregulated. The result is that success is now put down to the innate capacity for self-discipline and visionary self-awareness in some individuals, as opposed to others.

In 1973, Steve Jobs dropped out after one semester at Reed College. He 'then spent the next year learning the I Ching -a Chinese system of symbols used to find order in chance events - while dropping acid and dropping in on Reed's philosophy classes' (Newman). It was this binary system developed in China in 800 BCE that had provided the essential background to the digital revolution that would bring him international acclaim. This same binary system would revolutionise the work being done by Leibniz in philosophy and mathematics some five millennia previous to Jobs' period of discovery. Through exposure to the I Ching, Leibniz was able to discern that even the most complex aspect of reality could potentially be represented in binary form. Centuries later binary code makes up what we now commonly experience as 'digital realism'.

The I Ching was a system of divination that involved zeros and ones, lines and breaks, to calculate answers on the level of one's personal fate but also of the greater environment in which these changes would arise. Such probabilities were at once invisible and systemic within the universe. 'The 64 hexagrams of the I-Ching claim to represent nothing less than the archetypal situations of human life itself' (Walter). Jobs' fascination with the I Ching in 1974 would not have been unusual because, by 'the mid-1970s the counterculture had become the culture'

(Oppenheimer 6). Similarly, it could be argued that Jobs' adolescence spent in San Francisco was directly influenced by the development of two distinct overlapping political movements that were emerging within his immediate vicinity in the late 1960s; the New Left and the counterculture. Both movements were predominated by the appearance of white middle-class university students who believed that social movements were the key to transforming America's postwar society. While both communities based, the counterculture was focussed on reshaping the affective economy as a means of political action. For this group, the new communalists, the answer to society's ills laid 'not in changing a political regime but in changing the consciousness of individuals' (493). What this movement countered specifically was the centralisation of power throughout the cold war cerebral forms the clock and the mainframe computer, seeking to install in its place a reliance on the co-production of reality through the harvesting of a multiplicity of informational flows.

Mapping Over Power

From his earliest successes, Fuller was consumed by a passion to join the reconceptualisation of the global with the practicalities of design. His creation in 1942 of the Dymaxion Map made a visceral connection between the local and the global by representing the earth's surface as a series of parts each related to the whole and connected by great circles, signifying an interdependent, spherical system held together by the twinned forces of tension and compression. This would form the perfect conceptual blueprint for an apprehension of the world as fundamentally interconnected; a worldview we now, as a public, readily identify with. This was in sharp contrast to the map Fuller took as his starting point. Fuller's map promised to deliver a comprehensive view of the Earth's surface, free from the projections and distortions of the Mercator map that were as much practical as they were political in nature. In Fuller's estimation, this map was now an artefact, an example of an out of date technology no longer fit for the second half of the twentieth century, which held the potential for innovations for travel and navigation heretofore unimaginable. These innovations would fundamentally alter the economic geography of the planet.

In its first incarnation, the geodesic dome was designed as a portable shelter capable of supporting and expanding the Cold War surveillance systems of the burgeoning military-industrial complex. The 'geodesic domes were actually installed worldwide as the enclosure of choice along the United Distant Early Warning electronic warfare front. The spherical form of the domes provided maximum structural transparency for radar devices scanning the Arctic sky for incoming Soviet missiles, and the ability to fly these domes into remote, often mountainous regions suited the concept perfectly' (Leslie 166-167). The essential components of these domes were attractive to the United States Army because they could be air-dropped via helicopter and assembled quickly by infantry in a combat situation. Fuller would later transfer this delivery technique into the civilian sphere, in his project for creating aerial zones of urban renewal evident in his "Skyline for Harlem" project.

Fuller's military background placed him at the forefront of the transition of a formerly industrialised society in the World War II era to one increasingly driven by advances in computerisation. This transition profoundly recast humanity's conception of itself and its supporting ecology. Thomas W. Leslie reasons that Fuller's 'willingness to envision the planet as a collection of dispersed data points connected to, but not fully represented by, the shipping and transport links of his cartographic projects', enabled him to eventually join his 'vision of a geodesic world' to an expanding infrastructure of capital (167). This infrastructure was one based around a 'non-hierarchical, ultimately efficient network of distribution for economic and physical resources' (Leslie 167). Its design married well to the ambitions of the countercultural movement of the 1960s.

Its adherents enthusiastically endorsed Fuller's ambition to actualise a universally scaled consciousness founded upon an individual's singular perception of locality. The negative aspect of this actualisation recalls the ultra-efficient mesh of global colonisation initially assembled during the Mercator era. Its hierarchical organization of the globe displays powerfully an interdependence and independence of individuals within a global system, the ends of which concentrated power toward the hands of a minuscule elite. Therefore, in judging the countercultural movement of the 1960s, one would do well to recall that these advances acted to the benefit of the world's European population as a whole. European imperial power became the standard for civilisation over the course of five millennia through the espousal of a socio-objective measure of performance. Selective debates over technology and society were conducted amidst a backdrop of implicit global structural and productive inequalities. From such a perspective, resources around the globe were justified in taxation and the continuity of poverty to wealth dissimulated.

Within this distorted structure, it is difficult, if not impossible to transform the enduring infrastructure of empire into a pioneering network capable of supporting the greater whole of humanity. Fuller's Dymaxion projections are a testament to how much his vision of Spaceship Earth as a system of ephemeral networks is at once indebted to this cartographic legacy, and riddled with a cybernetic scientific abstraction 'intended to be employed towards social, distinctly anti-political ends' (Leslie 168). Fuller's model is ultimately one built from an assumption of scarcity, as is his 'considered use of technological innovation as a method for distributing or conserving scarce resources' (168). That it now finds affinity with twenty-first-century concerns around sustainability and digitisation does little to mitigate the reality that as a model of global projection, it remains fully divorced from its role in perpetuating white imperialism, and as such continues to play its role in promoting ingenuity as the world's cognitive energetic through line. Thusly it perpetuated the myth that certain technologies were 'intrinsically radical and could change the world simply by existing' (Kirk 388).

This myth would eventually become capitalised upon in the products made by the counterculture's movement of thinkers. They sought to capitalise on an image of the counterculture as an alternative politics where, in sensory concert with technology, individuals could simultaneously transform themselves as well as their relationships within society. They were intent on building nothing short of a new intuitive environment infused with underlying environmental principles borrowed from the transcendentalism of the late Victorian era. The talismans of this movement would be NASA's photographic image of the Earth from space and the yellow smiley face, (a precursor of the now ubiquitous emoji). This movement was fundamentally conservative and reactionary, referencing the cold war seminal text, 'Norbert Wiener's Cybernetics as well as the Wall Street Journal', and in many respects going right of them towards an embrace of full-scale libertarianism (Turner "The WELL" 495). Community, in their minds ideally was self-ruled, transpersonal, and technologically mediated. Implicitly it was also about the recovery of a white Euro-American culture that participants feared was being lost as a distinct identity which could be re-enlivened through platforms that facilitated virtual tribalism.

The imperative of this movement can be found now in online forums such as Reddit and 4chan, who also subscribe to this concept of recovering a diminished white identity through the establishment of strong interpersonal networks. These attitudes substantially predated the rise of computer networks, and yet remain evident in the colonisation of specific social media platforms and the deployment of decentralised networks of authority amongst those in the so-called "alt-right". Similar to the late 1960s, these forums are heavily authored and populated by a largely upper-middle-class, college-educated white male demographic who are highly skilled and have access to some form of gainful employment.

Few early adopters of these virtual platforms were consciously aware that through their engagement and contribution they had commodified themselves and allowed their interior logics and torments to become the property of corporations. The self in this sense was becoming a commodity that was being sold to other commodities to be consumed as entertainment, and to furnish these corporate entities with an endless source of personal data to mine and repurpose for profit. This would become overtime the ultimate form of recycling in a virtual world increasingly polluted by social polarisation, graphic violence, and fake news. As virtual ecology becomes increasingly precarious so too do social and economic opportunities, a disease infecting the prospects of even the top 10% of earners, who now like their 90% counterparts must face a world where people are moving from one job to another, one city to another, one social circle to another, and therefore are forced to find more and more novel ways to instrumentalise one another for survival. Within such an ecology communication cannot be seen as wasteful. The reckless wastefulness of liberal economics which fuelled the cowboy economy of America's founding century would be forced to give way to the efficiency required for the neoliberal economics of Spaceship Earth. Indeed, it was the astronaut economy of the Cold War era which necessitated the making of markets that were an imitation of a time when life itself was increasingly perceived as unique, precious and finite.

The cybernetic principle of viewing systems essentially as information revolutionized the theory and practice of both the economic and earth sciences in the second half of the twentieth century. Cybernetics introduced new concepts such as the phenomenon of 'feedback' in market and ecological systems and stimulated the university and private industry to perpetually develop new tools for applying those concepts, in the form of computer modelling and the globalisation of information. Ultimately, it had the effect of transforming the concept of the Earth to conform with human specifications. Moreover, according to Rob Coley, 'the coming of the Anthropocene asks us to imagine the earth itself as an image-recording medium, as that upon which a geological graphé of human activity has been inscribed' (50-51). This quality is fundamentally inhuman or posthuman, insofar as the human world has become synonymous with data. Data is what both inscribes and prescribes a human existence that is now subject to ceaseless tracking, documenting, and modulating and where reality itself is founded on the basis of such data. Humanity, and by extension mankind finds itself in a subordinate position to inhuman processes that busy themselves with tabulating, indexing and sorting the entangled relations that makeup power relations on this planet and, in so doing, dominate the world.

The Anthropocene in this scenario becomes then nothing more than an affect of computation. Though perhaps more significantly, it is precisely for that reason that it is capable of bringing to an end the very notion of a singular world. At a time when the human is confronted with the likes of global warming, and climate change more generally, those with the greatest cultural capital on the planet refuse to acknowledge that these phenomena are massively distributed in time and space relative to that distribution of power. Ironically, the cost of such wanton neglect can be calculated but not witnessed. Such violations of the so-called laws of nature can be experienced but not represented, and this is precisely the case because such processes reveal indissoluble connections between white supremacy and the uneven agencies of peoples of colour that is the enduring legacy of colonialism in the spaces and times that exceed its formal manifestation. The crisis of the Anthropocene is not that the world is ending, but that a world may be ending through the introduction of the non-human into planetary consideration and therein competition for resources on this planet and indeed beyond it. This, therefore, can be read as a crisis of mediation, where glitches are now being generated that reveal that human perception itself may be subject to a horizon line, and the deep implication of this as a means of navigating the universe no longer confined to an obvious definition of an 'us'.

The irony of this situation is that what is nonhuman may one day outsmart us, expressing to us our world, and in so doing force us to perceive the world otherwise through processes that require an act of attunement we have heretofore shunned in others. Such others have already refined an attunement to the inhuman because it has always already been an immanent part of their experience of being, and therefore being subject to perpetual instabilities visited upon them by others. Perhaps we are not at the end of the future after all, but instead, being compelled to think of ourselves as viscerally elsewhere..

Chapter 5
Coda: The Necessity of Reattunement

Malign Vibrations

Perhaps it is time, as cultural critics as diverse as Walter Benjamin and Sun Ra have been telling us for years, 'to think about sound as an instrument of oppositional consciousness, particularly in relation to struggles over space' (Wald 673). It was easy for a late 1960s America to embrace 'the Beach Boys' vibrations,' which were readily 'absorbed into popular culture as the sound of white California' (Wald 674). Still, there was a disturbing undercurrent within this vibration. In the summer of 1968, the Beach Boys recorded a song that was authored, in part, by Charles Manson. A B-side to the single "Bluebirds Over the Mountain," Manson's track was originally titled, "Cease to Exist." This track was never publically credited to Manson, but rather to the Beach Boys' drummer Dennis Wilson. Wilson had befriended the Manson family that summer taking them into his home to experiment with sex, drugs, and violence, and briefly into the space of his musical career (Dowd). The lyrics of "Cease to Exist" (later retitled by the Beach Boys as "Never Learn Not to Love") implored its implied female listeners to 'give up your world, come on and be with me," and "submission is a gift, give it to your lover," clearly promoting Manson family values. When the single was released these were readily absorbed by the Beach Boys legion of young fans as an idealisation of learned female dependency ("Never Learn"). The friendship between Wilson and Manson would end shortly thereafter when Manson flew into a rage when he came to realise that he was not credited with co-writing this track. Prior to that situation, Manson began to openly share his predictions of an upcoming race war with Wilson and erupted into regular fits of rage in front of him. Things came to a head when he threatened the lives of Wilson's children as revenge for the Beach Boy's failure to credit him. Wilson felt so in danger that he abandoned his own home, quickly giving it over to Manson as a way to quell his fury.

Later that same year, Manson became obsessed with another band promising to deliver on an emblematic whiteness; The Beatles. Zahara Hill elucidates the relationship between Manson's avowed racism and his white supremacist interpretation of the Beatle's "White Album":

> He believed the Beatles' 1968 album emanated with race-baiting undertones. Manson took inspiration from one song, in particular, "Helter Skelter", in titling his aspirations for the massive race war he hoped would result in the self-destruction of blacks. While the song appears to be a mere tribute to the complexities of love, Manson alleged he interpreted "Helter Skelter" and other songs on the album as a call to racially divisive action (Hill).

Manson became familiar with the Beatles' music starting in 1964, in the midst of his ten-year period of incarceration for a series of petty crimes. Manson listened compulsively to their "White Album" throughout the Spring of 1969 and was convinced that the lyrics of the track "Revolution 9" informed him of the exact date in which to enact a series of ritualistic murders. These murders would initially erroneously be attributed to the Black Panthers by an institutionally racist Los Angeles police force. The dubious evidence for this was based on the methodical scrawling of the Black Panthers' commonly known terminology in blood throughout the crime scene by members of the Manson family.

Manson believed the Watts riots in Los Angeles of 1965 were the prelude to a race war in America that would bring about a revolution. The war would be against the white establishment and their defensive forces. The military and police, "the Piggies," would act as the adversaries to a militant army of young blacks. Manson preached that there would be widespread death and destruction as a result of this conflict now soon at hand. 'Among his deception tactics was convincing [his white] followers that if they didn't commit the murders, Blacks would rule the nation and kill or enslave them' (Hill). Given the extreme tension in the country around race relations at this time, Manson was convinced that authorities would construe the murders he incited as the work of the "Blackbirds" who in accordance with his interpretation of the Beatles lyric, were now beginning to "arise" in one final, violent act of vengeance against their white "overlords."

Manson held the fantasy that when this war came to a close, blacks, having been in the desert of bondage so long and thus having developed no aptitudes through which to rule society, would embrace him as their chosen Messiah. This, only to be betrayed by him on a cosmic scale, establishing a new era of white supremacy on Earth led by him and his white female child army. Manson essentially believed that his cultural ascent was one that would take on interstellar proportion. Similar to the moon shot, it would function as a means of elevating humanity in the wake of a series of profoundly destructive events. Vietnam and Apollo 11 all rolled into one. Shortly after the murders, Manson visited Wilson's new home, 'telling him that he had "just been to the moon" and demanded money,' which Wilson quickly gave to him to provoke him to leave his property (Gaines 217).

Sonic Disorder

The Beatles' "White Album," released in November of 1968, attempted to give their listeners "the sense of a world moving beyond rational explanation"' (Orlando). Halfway through the album their listeners are immersed in a soundscape resonant of a world without order, where reality is quickly descending into pure chaos. 'The music seemed to absorb the panic and violence of 1968, a world that when interpreted musically, reflected its '"beauty, horror, surprise, chaos, order; and that is the world, ... created by, creating for, their age"' (Orlando). The music was meant to prefigure a near future where a revolution would come to pass. At the time, John Lennon claimed that the album's track, "Revolution 9" was 'an unconscious picture of what I actually think will happen when it happens, just like a drawing of a revolution"'(Grow). Lennon was attempting to create an atmosphere where listeners could viscerally experience a revolution in progress, which begins with recording an 'engineer testing the studio's No. 9 input and end[s] with what Charles Manson admiringly described as "the sounds of the end of the world"' (Orlando).

For over eight minutes the listener is lead through 'an avant-garde odyssey built with around 20 sound-effects loops, including samples of Sibelius' Seventh Symphony and part of the orchestral overdub from "A Day in the Life"' (Grow). "A Day in the Life, features in the lyric, 'see the worst thing about doing this, doing something like this, is I think that at first people sort of are a bit suspicious, "you know, come on, what are you up to?' ("A Day in the Life"). Lennon, unlike co-writer McCartney perhaps, believed that he was up to nothing short of creating '"the music of the future"' ("Revolution 9"). This precarious enterprise had to take into account both the nightmare of the present and the potential for a future of restored global unity in the world. At the time of Lennon and McCartney's experiment it seemed wholly apparent that humanity was heading towards a long good night full of rage and resentment. This situation could only be remedied by the world looking far outside itself into the cosmos for meaning. In December of 1968, Apollo 8, the second manned spaceflight mission flown in the United States Apollo space programme, was launched, becoming the first manned spacecraft to leave low Earth orbit, reach the Moon, orbit it, and return.

While in orbit the crew made a Christmas Eve television broadcast. During the broadcast, they read out the first ten verses from the Book of Genesis, the opening chapters of the Old Testament with its themes of creation, destruction and re-creation to a white civilisation that very much felt itself to be on the verge of dying out. Space held out the promise of its salvation and renewal. At this same time, Charles Manson was becoming wholly preoccupied with the other end of the Judo-Christian spectrum, the Book of Revelation, the final book of the New Testament which predicted the Apocalypse. This book sat amongst the touchstones that he returned to again and again to shape his ideological position. The others were Dale Carnegie's self-help book *How to Win Friends & Influence People,* which taught Manson techniques of friendly persuasion and social manipulation of those around him and L. Ron Hubbard's *Scientology: The Fundamentals of Thought*, which informed his belief in the importance of changing oneself, understanding the unity of humanity's destiny and appreciating than human life transcends beyond its time on Earth. Manson would go on to re-interpret these ideas to reinforce his own doctrines and to maintain influence over his followers. Manson believed The Beatles were the final element of this grouping through which he could achieve his own status as a cultural icon and conceptual guru within the burgeoning counterculture. Individuals involved in this movement were driven to adopt new ideas, new lifestyles, and, new leaders as a means of bringing about a revolution in values and reinventing Western culture itself. The lyric of Revolution belies the counterculture's allegiance with a libertarian versus communitarian stance, 'but if you go carrying pictures of Chairman Mao, you ain't going to make it with anyone anyhow' ("The Beatles - Revolution Lyrics") The road to revolution in the sense is paved with another type of economy altogether.

Manson, 'used his powers of persuasion to make his followers believe "that all we had to do was ask the universe for what we wanted and it would be presented," Dianne Lake, a "family" member who didn't participate the murders, wrote in a recent memoir. "In the connection with Dennis Wilson, it appeared that was precisely what had happened: Charlie had led us to the communal promised land — everything he'd asked for had come to pass"' (Rosenwald). Lake recounts Wilson driving Manson and the girls to the back of a local grocery store in his Rolls-Royce, where they showed him how the family had been able to subsist by salvaging edible food from the store's dumpsters. Wilson was literally seduced into believing in the Manson's family's Freegan value system by Lake:

> We all laughed and sang all the way to the Dumpster, dragging Dennis by his hand. The best thing we found on this run was a flat of strawberries. After culling out the bad ones, we had enough to make him a strawberry cake complete with fresh Cool Whip. Charlie was leaning against the Rolls watching as we showed Dennis how it was done. "Dennis, do you know how much good food is thrown out in America?" he shouted. One of the girls popped a fresh strawberry into Dennis's mouth and we all hopped back in the car. That night the girls and I made an entire meal with the produce and other discarded food. Then we presented Dennis with his cake…Such joy (Lake qtd. in Rosenwald).

Manson again lets his young female followers do all the work while he sits back and casually riffs on the decadence of American culture. There is a sensual sexual valance attached this message that resonates with an impressionable Wilson, who readily comes to associate the satiation of his personal desires with a greater set of social freedoms. Manson's cult leader persona adhered perfectly to the libertarian values of California's 1960s counterculture which celebrated a lack of personal responsibility in light of a larger goal. This value system was informed by a heady mix of sexual promiscuity and excessive drug use, weak personal ties, casual interpersonal relationships, and a strong yearning to interrogate and disrupt societal norms. This resonates with The Beatles avowed belief system as recounted by John Lennon in a 1980 interview with *Playboy* magazine. Lennon asserts that, "'The whole Beatle idea was to do what you want, right? To take your own responsibility.'" ("John Lennon Interview.").

In that same *Playboy* interview, Lennon chastises their fans who mourn the dissolution of the band as the passing of a kind of sacred era. Lennon, extending this metaphor of a type of religious awakening of their fans to conclude that those who are still looking to them for answers are a deluded bunch,

> If they didn't understand the Beatles and the Sixties then, what the fuck could we do for them now? Do we have to divide the fish and the loaves for the multitudes again? Do we have to get crucified again? Do we have to do the walking on water again because a whole pile of dummies didn't see it the first time, or didn't believe it when they saw it? You know, that's what they're asking: 'Get off the cross. I didn't understand the first bit yet. Can you do that again?' No way. You can never go home. It doesn't exist. ("John Lennon Interview").

Two days after this interview was published, Lennon was shot dead on the street in front of his New York apartment by an American evangelical Christian, Mark David Chapman. His controversial murder was given over to the public as the final unorthodox iteration of the Beatles as a mythically inspired rock band. The Beatles had once been countercultural Messiahs for a wayward generation. After Lennon's murder, there would be no obvious route to salvation in what was slated to come, no resurrection of The Beatles sound, nor further instructions on how to consciously rise given by them. There would be no homecoming for the civilisation that had to reckon with yet another murderous historical passage. The world had moved on to what surely would be its next.

The Sun Also Rises

All of this could be traced back to a founding act of civilizational violence, be it the Crucifixion of Jesus on the cross, or in a more contemporary sense, the formation of the Middle Passage out of Africa. These revolutionary events forced culture to become immediately mnemonic. Those exiled by these developments salvaged various cultural devices, patterns, ideas, and associations to aid the process of remembering something of what had been left behind in the wake of these revolutionary events. This enabled them to take something from the ruins of their experience, even as the actions surrounding these events themselves became quickly dematerialised, passing into the realms of mythology. In the wake of such profound acts of foundational trauma, culture becomes very much about memory re-materialised externally through the labour of production and the force of enunciation. The same founding trauma has to be passed on and re-articulated for each generation in order to sustain it.

Within the confines of this mnemonic economy, language becomes a virus and as such, there is a whole strain devoted to the alienation of the body by the invasion of outside forces. This alienation expresses itself directly in the white somatic vernacular, versus the black one. For Kodwo Eshun, 'there's the key thing which drew me into all this: the idea of alien abduction, the idea of slavery as an alien abduction which means that we've all been living in an alien-nation since the 18th century' (193). Eshun is referring to, 'the mutation of African male and female slaves in the 18th century into what became negro, and into the entire series of humans that were designed in America. That whole process, the key thing behind it all is that in America none of these humans were designated human' (193). The virus in this instance became the form, and the music that was produced by African Americans heretofore reflected their status as the non-human part of America's visceral composition, as something always out of sync with its operational protocols. The symptoms of which in the 1960s were protest and riot, tools in many senses left over from the Enlightenment definition of rank and rebellion.

The African-American body in these schemes was always perceived by white America as in some profound sense treasonous. This judgment involved the sentence of guilt to be laid upon the black body for the crime of betraying one's country by virtue of their non-being. Thusly, these former Africans were denied their American citizenship. In a political sense, they became pointless beings further diminished by their perceived incapacity to find their way home. The only solution for African Americans in their self-defence was to name another place home. Sun Ra does this by saying that he is a citizen of Saturn. This itself is a treacherous act insofar as it implies his having passed through a territorial border without any white person's permission and under no white person's authority. Rather, Ra is serving at the discretion and indeed pleasure, of an alternative sovereignty.

Ra makes a remarkable crossing into the cosmic dimension of sound, producing 'the music of the electromagnetic field, the music of radio transmissions, ... of electrical disturbances, the atmospheric cosmic disturbances that exist in the sky' (Eshun 184). Ra made of himself and his Arkestra a technological apparatus. Acting as scientists they synthetically produced the 'molecules of a new people' that 'may be planted here or there' in worlds to come (184). Out of this material they are producing 'the new astro-black American of the 70s' (184). These are folks generated with no sense of fidelity to what has gone before, but as a novel generation poised at the threshold of new environments. They strive to become a new sensory life form; one able to escape the Earth's historical bonds through the medium of music. Through a vibe and vibration that most outsiders considered impossible to achieve they are transporting themselves out of ancient Egypt into outer space, bypassing along the way the suffering of the ages.

Firing at Acoustical Range

Gayle Wald explains, 'Ra's vibrations' resisted… incorporation-
and not just because, like [Brian] Wilson [of the Beach Boys], Ra was
viewed by sceptics as more than a little unbalanced himself' but rather
because Ra positioned himself explicitly 'as a medium of consciousness'
(674). If the Beach Boys "Good Vibrations" was a miniature symphony
composed along the lines of a song about a guy picking up vibrations from
a girl, Sun Ra's work was, by contrast, composed along the lines of a song
about a whole race of people picking up vibrations from a universe. This
difference in proportional aim was so because for Ra vibrations were 'the
conduits to new worlds,' not merely a way for some folks to feel accepted
within this one (674). Ra was a student of acoustics throughout his life. He
was convinced that sound literally had the ability to touch the listening
body and to transport it well beyond the proscribed realms of rationality
that modern science had discursively assigned to it. Ra was convinced
these predominantly white male Western cyber-technicians were 'unable
to compass music's utopian potential beyond physical "laws,"' and
therefore unable to locate the "sacred arena" of vibrational space which
had the power to [both] produce new possibilities of social attunement and
new modes of living' (674). Ra's music was the first to attempt to insert
blacks into the Space Age at a time when African Americans working in
the space industry were hidden from public view and the space race
codified as a white only affair; American and Russian.

Ra held an innate distrust of Christianity, blaming it for never
acknowledging its role in perpetuating slavery, and for instilling a false
consciousness of the mythic past in the minds of the black populations
made object to it. 'Ra was no less opposed to the mythic future that the
Christian religion offered black people' (Lock 32). In opposition to this
'he extended his myth world back to the ancient glories of Egypt rather
than to old testament Patriarchs and upward toward a heaven that he
depicted not as the location of a promised afterlife, but in the light of
astronomy and modern technology, as a real, living future in outer space'
(34). In the film, *Space is the Place* (1972) Ra, as the main protagonist is
portrayed as a time traveller visiting Earth via a spaceship relaying himself
between Chicago in 1943 and Oakland in 1972. Ra's mission is to recruit

those dwelling within these localised African American communities to join him for a life in space. He is there specifically to help them escape from the corruptions of a mainstream media network that largely excludes black prophetic voices and the United States government which disproportionally visits surveillance and violence upon black communities. In one particularly poignant scene set in Oakland in 1972, viewers witness Ra being kidnapped by the FBI and 'sonically' tortured 'with a recording of the Confederate anthem "Dixie"' (Zuberi 991). He is rescued by three black youths, but shortly thereafter at a concert in which he and his Arkestra perform, the FBI return and attempt to assassinate Ra. The youths once again come to Ra's aid. They are subsequently rewarded with transportation to outer space to live in harmony with Ra and his colonial brethren.

Ra's acts of redemption in this narrative are fundamentally performative in nature. He seeking to address an audience of blacks whose experience of both psychological and technological breakdown exists a consequence of a liberal capitalistic society in which they were often perceived as broken instruments. When that history was redressed, however, it would be possible for them to express themselves through a specifically black vernacular, redirecting a language around creation and use that enlivens agency. This is a language that is ideally capable of addressing their long history where blacks were construed *as* materiality and one enabling blacks to reconceptualise race itself as an artefact, as well as, a symbol. Through the use of such language, it would become possible for blacks to recognise their unique affinity with cold war technologies including space technology, portable cameras and recording devices, nuclear power, and the internet. At the same time, Ra believed it was crucial for blacks to associate their history not just with slavery in the Americas, but rather with ancient Egypt, because, it immediately reconnected them with 'an African-American heritage rooted in a civilization known for its technology. In Sun Ra's formulation, the historical role that blacks played in technical achievement could serve as the foundation for the development of future forms of black-valued technology and the basis for African American claims on the landscapes of the future' (Kreiss 61). This message was crucial at a time when the

Black Panthers were telling their followers to be wary of space exploration
as yet another derivation of white coloniality.

In January of 1969, six months prior to the moon landing, a
drawing was published in *The Black Panther* magazine entitled
"'Imperialist Plans...'" which depicted pigs (white exploiters) dressed as
policemen driving black slaves on an outer space landscape. One of the
pigs is saying "hey handle those slaves with care we're gonna need them
for Mars, Pluto and all the other planets". A slave is responding: "I knew
we should have stop [sic] this shit before it got off the ground"' (Douglas
qtd. in Kreiss 61). The Panther's focus was on direct political intervention.
They aimed to exert control over these emerging space age technologies
through their cultural appropriation by oppressed communities of colour
worldwide. Theirs was essentially a Marxist strategy calling for people of
colour to seize the means of production and thus alter the course of history
to tilt in their favour. Ra was very much less materialistic in his
revolutionary aims. Indeed, he wanted to alter consciousness at a much
deeper, anti-materialistic level by redirecting the positive energy of black
communities towards the empowerment of a completely alternative
society. This would function as a means of re-authoring their cultural
trajectory and in so doing, help them find new ways to peacefully
commune with technology. In 1968 he began to extend his mission to
California and beyond to the United Kingdom, Europe, and finally Egypt
in 1971.

Upon returning to the United States, Ra was appointed as a
lecturer at the University of California, Berkeley in 1971, teaching a
course entitled *The Black Man in the Cosmos*. It was an unorthodox
association for which he was 'never paid' (Walsh). Few of Berkeley's
matriculated 'students took the course, although large numbers of Oakland
residents attended. The reading list included the Egyptian Book of the
Dead, the Bible, and books on hieroglyphics. Ra also spent part of each
class playing the keyboards' (Walsh). Few students understood the
implications of what Ra was transmitting to them:

> "I'm a spirit master," he said. "I've been to a zone where
> there is no air, no light, no sound, no life, no death,
> nothing. There's five billion people on this planet, all out

of tune. I've got to raise their consciousness, tell them about the wonderful potential to bypass death" (Sun Ra qtd. in Walsh).

The mission was to get black folks to settle within mythocracies as opposed to remaining in the theocracies, democracies, and plutocracies that never were designed to elevate or serve them. Ra's 'pronouncements were drenched in his unique cosmic mysticism, and his band members claimed he had telepathic powers. Most importantly, he made music, as he wrote, "rushing forth like a fiery law"' (Walsh). It is significant to perhaps recall that Ra briefly moved to Oakland in 1971 'at the invitation of Bobby Seale' (Bengal). Ra and the Arkestra lived in a house owned by the Black Panthers, whose military uniforms reclaimed symbols in a way that paralleled Sun Ra's mythic dress, both offering their own version of Black Power: the Panthers in their black berets and leather jackets, Sun Ra in his moon boots' (Bengal). Both were prototypical figures of a blackness that was heretofore unrealised, whose appearance shattered the peace of a white supremacist hierarchy and brought into being resonances of another reality that now wished to suddenly, loudly shout out its appearance through figures that were never meant to exist in this society. They were not interested in racial equality insofar as it connoted a self-sameness with something the never were, and indeed never could be, within the ideal of America's nationhood.

What they meant to do instead was occupy a neighbourhood, a place were minor histories occurred and where minor keys could be played out to block out the negativity of the world at least temporarily, or so the myth goes, to keep the dream alive. Here the present could be given its place amongst other ancestral objects on the mantle of a different sort of being; one capable, finally, of otherwise illuminating consciousness. Neither the Black Panthers, nor Sun Ra wanted to be associated with what they saw as a failed species, or included in a definition of humanity that equated itself with an apocalyptic-seeking whiteness. Their being would, therefore, by definition, have to run parallel and opposite to the one assigned to them. For both groups mobilising blackness itself was the ultimate destination of release and transcendence from the death and destruction a white humanity had visited upon the Earth. Redemption would necessarily have to come from elsewhere, a space outside of all of that foretelling, derived from the unknown regions of this planet's history where a different sort of life sense was and could be enacted freely. The Black Panther movement was officially disbanded in 1982. Sun Ra left this planet in 1993. Neither party would materially persist to meet the arrival of a new millennium and perhaps the end of the future as we have come to know it.

Agency within this context cannot be reliably achieved by continuing to adhere to vision as the power regime of this planet, but rather in order to imagine alternative futures one must now make recourse to sound. Peoples of colour have long been aware that for centuries now vision has operated as a master sense, one that is particularly reserved for predicaments that were not part of their worlds alone, but rather transcended the boundaries of the known world to destabilise their indigenous codes of trust, expression, and communication. Vision as a medium had worked to still their power specifically within the paradigms of racism, the state, and institutions that represented national culture. Going back at least to the era of American slavery and forward to the twenty-first-century era of Black Lives Matter, it is possible to recollect the use of sound, in concert with spectacle and code, to engage with the phenomenal nature of this planet, and pitch its vibrations for the betterment of living in continuum with other entities - living and non-living.

Benjamin asserted, 'what characterizes revolutionary classes at their moment of action is the awareness that they are about to make the continuum of history explode' allowing for another kind of consciousness to come into its own (395). 'Sun Ra called his band the Arkestra-part ark, part orchestra-and defined its mission as more religious than musical' (Staples). Theirs was always going to be an instant of launching, versus arrival. Ra was preparing them to outlive culture if need be. Their goal was not to exist but rather to persist; not in the form of materiality, but rather in the form of myth. Ra advised his followers that '"myth speaks of the impossible, of immortality. And since everything that's possible has been tried, we need to try the impossible"' (Ra qtd. in Staples). Perhaps Ra was just waiting out the storm, allowing Earth's stranded whites to tire of pursuing him and his racial brethren across the ages. Black culture would ultimately triumph by slipping the bonds of mortality first instilled upon it by mass enslavement and by essentially outliving the materialistic belief systems that perpetuated its association with capture and limitation. For Ra, 'reality equals death, because everything which is real has a beginning and an end' (Ra qtd. in Staples).

The Dying of The Light

What in the new millennium seemed a new reality to white Americans and Europeans, however, had been the condition of life for the planets subjected peoples now for some five centuries. Theirs was a world made devoid 'of liberal freedom, reflection, consumption, rights and critical distance' that their white counterparts took for granted as universal properties assigned to human subjectivity (Colebrook 103). It was they who set these conditions for the 'world' and they who enacted a proprietary relationship to the world relied upon to produce 'leisured time, reflective reading, democratic debate, rights and reason' (103). Seldom did this world include an explicit recognition that its persistence was based upon the ending of another world through its rapacious extraction of life and energy. The end of the future, therefore, can be said to be a dawning of a condition where these worlds are finally allowed to merge to produce a world in which 'order, abundance, personhood or leisure' are no longer given attributes to those dwelling within the American and European imperial axis (Colebrook 103). The consequence of this has produced a great deal of fear, anxiety, anger and depression as it becomes evident to that the future of 'humanity' has been suddenly profoundly 'darkened'.

Evidence of this crisis of white privilege can be found at work in 'Oxford University's Future of Humanity Institute', which 'sets itself the task of guarding against the loss of thinking, intelligence and technological maturity, as though these predicates were definitive of the human' (Colebrook 103). Claire Colebrook wryly remarks, 'a life without this Oxonian humanity would, it seems, be the end of the world' (103). This loss of humanity, (read Euro-American privilege) stems from our melancholic idea that whites in fact manufactured this world to begin with through liberal mechanisms of justice, fairness, tolerance, and above all a fundamental claim to selfhood. What the post-apocalyptic reading of the future points to i.e. the Anthropocene, is thus a depiction of the world as it has been for the majority of humans fundamentally denied these privileges. It was they whose position in the world has always already run counter to the assumed position of the affluent Euro-American white male. What has served the latter position has been a global geography mapped along lines of monopoly, oligarchy, lineage, plutocracy, waste, and wreckage. What has been put before the majority of us as a salve to these ills is hyper-mediation and its mainstay hyper-consumption. Those recently denied or made devoid of their personhood in the developed world are in sum directed to these new worlds –whether they be online or in outer space, where once again endless expansion is valorised with no thought paid to the depleted, volatile, exposed and rigid world right outside the door.

This situation forces those who occupy the Anglosphere to examine how the interior formation of their world is predicated on the externalised destruction of Others. If a greater proportion of the Anthropocene is acknowledged as the responsibility of Western nations, so too must be acknowledged that this rush towards responsibility equally intensifies the perception that those that occupied these privileged lands also remain in the exceptional position of being able to make their own history thus continuing to define its universalist terms. Despite the fact that their planetary predominance spans no more than a few centuries, they remain convinced that it continues to entitle them to promote democratic, market-based solutions globally that will save 'the human race'. It is this entitled class that has become so adept at extending its global reach that it can now scarcely recognise the concept of a limit to its ability to organise and perpetuate conquest having born almost nothing whatever of its costs as compared to its worldly Others. Their narrow grasp can hardly be relied upon as a true arbiter of how in the future we might come to recognise humanity on this planet. There dominant units of measure which long persisted along the axes of property and propriety are now being realigned to include within them valuation and quantification arching towards a futurity where nothing is beyond measure.

We have lived through millennia of humanity in which time and energy have been seized, stolen, appropriated and co-opted in the drive to obtain greater profit, greater territory, and greater mastery to be reaped from this planet. The violence of that endeavour gets lost in time, lost to capitalist universality. In order to counteract this movement, we must resist falling in line with a recursive stance that demands that culture universally back-date reality to the West's experience of the post-WWII period. At the same time, we must strive to achieve distance from the tyranny of contemporary neoliberal commensurability. In doing so we must commit ourselves to seek out forms of experience that exist beyond the optimisation of reality and interrogate technology itself as an enduring infrastructure for controlling human subjectivity. It is through these skewed paths that humanity might discover in its place the asymmetrical articulations of sovereignty, the impositions of a techno-scientific elite, and the structural violence of institutional protocols throughout the globe that subtends its seemingly dire conditioning.

The Necessity of Reattunement picks up from Chapter 4, to explore the connection between the popular culture of the 1960s and the counterculture. Exploring the music of the Beach Boys, The Beatles and Sun Ra, it addresses sound as an instrument of oppositional consciousness, particularly in relation to struggles over space. Cultural phenomenon unique to that period including, the Apollo space missions, Charles Manson, the Black Panthers, the Beatles "White Album" unite in the cause of seeking salvation and renewal amidst a backdrop of social chaos. Each in their own way emerges as a key formation related to the counterculture in its quest to join the satiation of personal desires to a greater project of expanding social freedoms. The libertarian values of 1960s counterculture celebrated a lack of personal responsibility in light of a larger goal. This value system was informed by a heady mix of sexual promiscuity and excessive drug use, weak personal ties, casual interpersonal relationships, and a strong yearning to interrogate and disrupt societal norms. Popular music was seen as a conduit to promote these values. At the same time, vibration held greater political connotation as a means of extending understanding outwards. Through sound broadcast society found revolutionary ways to promote and sustain deep political introspection.

This probing yielded both great violence and a greater cultural reckoning. Walter Benjamin asserted, that what characterizes revolutionary classes at their moment of action is the awareness that they are about to make the continuum of history explode allowing for another kind of consciousness to come into its own.

Epilogue: The Futures of Ends

The End of the Future might be just around the corner, or more appropriately perhaps its ends. We are told repeatedly indeed how this all ends; in climatic disaster, as slaves to robots who have mastered the human race, or as one of those planetary millions on life support furnished by the benevolent aid of a basic income. Presently we are told, life could be better, while presumably, a large percentage are fairly willing to admit it is worse. Or at the very least hard work. Or to take the argument further, that we ourselves are being parcelled out into bits of data to be bought and sold, while all the while we are conveniently bilked out of our profitable share of its value. Such an equation has left we as humans altogether sensing that on many levels we are poorer for the effort.

The narrative of the future is one where desire and mourning comingle in order to sum up a world that is now in many ways perceived as subtracted from itself. The signs of life we had come to rely upon to register existence are no longer held within the liberal orbits of capitalism, imperialism, colonisation and slavery. Neoliberal life, bereft of these gravitational forces is perceived as ungovernable. Life as it is, is left to grapple with its remnant forms: indigeneity and nomadism. Humanity becomes its own artefact, its own object of contemplation, up there with the endless fascination for the dinosaurs. In the aftermath, we hunger for likeness and thirst for tribe. Like the meteor that killed the dinosaurs, we perceive climatic disaster as something that was unwittingly visited upon us by forces beyond our understanding. We also imagine it is something we still have time to outpace, to regress or to conquer through technology, in the way we have waged all wars – that is to say, against a common enemy. Within this scenario, 'it becomes impossible to think of extinction in anything but a narcissistic manner' (Colebrook).

The ending is all about us. Except that it isn't. What is threatened with extinction is not humanity, but the parts of humanity that were able to hold mastery over others. What will be taken away from us is the ability to exclude and diminish others. Claire Colebrook identifies this phenomenon as 'self-extinction' ("The Future"). This term holds value insofar as it only applies to those formerly assigned selfhood, or be recognized in relational manner to it. In the twenty-first century, this former self is forced to adapt to a new objectivity where in future it might have to adapt to the meagre living standards of 'slaves, the colonized, refugees, or indentured laborers' in order to survive (Colebrook "The Future"). What has been made extinct on this planet however is not resource, but rather potential. There will be no triumphing over adversity, no reason that will allow capitalism to locate the standing reserve and opportunity it once so easily lent to the figure of nature. Conquest was once the stuff of enhancement but has now ceased to be anything other than a situation which brings us into an uncomfortable proximity with the others we objectified along the way, whose aptitude for productivity we once classed as their virtue. The old formulas don't seem to work anymore and the only people we have to turn to for guidance it would seem are those we already have driven out of this world. Perhaps it is only them that are now available for redemption in our anticipation of what is to come for all of us.

Construction Ahead

Humanity, was, of course, never a generalizable concept and for that reason alone, it is neither advantageous nor desirable for it to be so as we come to the end of the future. Surely now new forms of barbarism will have their day in the sun. But what if we were to imagine a future that looked nothing like ours and that the end would come to us as nothing more threatening than the evolution of neoliberalism and the retooling of capitalism. What if we started from the point of a different ending. It could well be the case as Walter Benjamin recognised that for nearly the past century, 'mankind has been preparing to outlive culture if need be', and that it has been improvising out of the stuff found within its own body, merged with technology, the miraculous means with which to keep the end of its existence perpetually deferred ("Experience and Poverty" 733). In this place, 'everyone has to adapt –beginning a new and with few resources' (735). Rather than perceive this as a cause for melancholia the extinction of humanity could now be rewritten as a source of amusement (735). Paul Scheerbart was one such author, whose predictions for humanity foreshadow those of Elon Musk.

Scheerbart was one of those figures whom Benjamin credits with proposing a future where 'ordinary French or English gentlemen of leisure travel around the cosmos in the most amazing vehicles' ("Experience and Poverty" 733). It is the same male leisured class who stand now to benefit from the technology solutions posed to mitigate the ravages of climate change. If these chosen few are to thrive in future, Scheerbart advises that they look to their 'telescopes', 'airplanes' and 'rockets' as the source material for rebuilding the human form itself to adapt to the novel conditions of future survival. Scheebart has every confidence that technology 'can transform human beings as they have been up until now into completely new, lovable, and interesting creatures' (733). The creaturely aspect of Scheebart's project reveals to us the necessity of humanity incorporating the animal- like into the reboot of its design. At the same time, the benevolent feeling Scheebart requires of this beast suggests that it will be a product of both social and biological engineering. In a similar vein, Scheerbart's creatures are equipped with the ability to speak a completely new language of an 'arbitrary constructed nature' (733). This language is fundamentally "inorganic" and so are 'the "people" it generates' (733). Essentially, what Scheerbart is pointing the way towards is the advent of artificially enhanced lifeform that is capable of both directing the best qualities of white privileged mankind and harnessing the best aspects of humankind's technology upwards into the stratosphere.

What these creatures possess is a "humanlikeness," rather than humanity. This is humanity 2.0. The new lifeforms Scheerbart imagines dwell in 'movable glass-covered' dwellings, that allow them to exhibit their advanced culture social transparency within a greater 'culture of glass' whose advent 'will transform humanity utterly' (734). Such designs have already been born out amongst the tech campuses that have emerged in Silicon Valley in this decade and their populations posed as the beta testing group for various prototypes of behavioural evolution. So here we are standing at the dawn of a new era of technologically enhanced creation. A place where humanlikeness – a core principle of humanism – finally meets its match in what might in be termed the 'dehumanised' era of planetary living on. Silicon Valley as a whole 'encourages its workers to see themselves in rationalist terms as a programming problem—as a pattern of behaviors and rules in a complex system that, if analyzed hard enough, can be tweaked and modified to perform optimally' (Williamson 282). Its larger aim is to create, or endorsing new institutions and practices' that generate a culture of self-programmable labour through a 'world that is equipped with the ability to rapidly retrain itself, and adapt to new tasks, new processes, and new sources of information' (Williamson 282). Silicon Valley aims to scale up the new cognitive manipulation models they are currently beta-testing on their workforce in order to eventually lay the groundwork for new forms of privatised governance wherein Silicon Valley becomes the remote epicentre of global institutions, practices, and policies.

Ben Williamson asserts such governance would take place 'at a distance via its headquarters and their networks of affiliations rather than through state centers of government in ways which are intended to reproduce its innovative and entrepreneurial capacities to sustain informational capitalism' (286). Such a vision is both interiorising and retrograde when it comes to its relationship to the cities and the lives it seeks to fundamentally reposition and re-programme. Within such a terrain the digital promotes an imaginary image of itself as a conduit of social emancipation. In reality, it tethers its user to a future of tribalised competition. Even as Silicon Valley drapes itself in a rhetoric of 'friendly reassurances that competitive meritocracy will naturally and inevitably build a cosmopolis of creativity, achievement, inclusion, equity, resilience, and sustainability' it remains the case, that those that go on to be its constructors are disproportionately young, white, and male. Such individuals already possess the luxuries of time, disposable income, and technological literacy. What they desire to appropriate next perhaps quite predictably is space. The maturation of cognitive capitalism puts into motion a sequence of events that will ultimately drive their appetites forward in a time of scarcity. As such time the prototypical constructor functions as a 'self-sufficient lone wolf, a 'figure who is able to reconcile the imperatives of self-reliance and individualism with the current social immobility and cultural atavism' (Pinto and Franke 2016, 29-30).

The metaphysical concept "I think, therefore I am'' becomes in Benjamin's century the corporate imploration to "THINK". In our century, it becomes the neoliberal imperative to "think differently". The world is not one fed anymore on experience, but rather one surfeited by information. The inhuman creatures that Scheerbart predicted in his century have matured in ours by feasting on the leftovers of both culture and humanity. There is no room for regression. The mouse, Mickey, followed by the computer one, Apple, are now virtually obsolete; transitory forms in the evolution that lead to contemporary mankind. Mankind now stands in the place of the mouse and must face the realisation that 'it is possible to have one's own arm, even one's body stolen' ("Mickey Mouse" 545). Picture the prehistorical mouse preparing itself to survive a major extinction event even as the dinosaurs die out around them, gnawing at their ribs. Now picture mankind doing the same to its Capitalocenic institutions. Only it's too late because their experiences also have now been seized. Their memories stored as raw material to build a new civilisation to replace what was formerly theirs.

Recomposing Malady

Dispossession has become the tool of accumulation and therein materialism has entered into the realm of cognition. In this space, 'holding onto to things has become the monopoly of a few important people, who, God knows, are no more human than the many; for the most part, they are more barbaric, but not in a good way' (Benjamin "Experience and Poverty" 735). Not content to appropriate the human, these few also aim to compel inhuman life to contribute to further to the cognitive and communicative endeavours on this planet. Together they form a new resource ecology composed of biological intelligence, artificial intelligence and computing algorithms animating competition and cooperation amongst humans, nonhuman animals, and multiagent systems.

This recognition of intelligence as existing within a plurality owes a great deal to neoliberalism's individualistic ideology and its prominence of place within the information technology industries. Within these industries, neurological variations considered outside the cognitive norm prevail amongst a disproportionate number of its workers. The glorification of their labour has run parallel with the rise of the neurodiversity social movement, which advocates on behalf of those diagnosed on the autism spectrum to have greater acceptance. The movement argues that their neurological conditions aren't medical disorders, deficits, or dysfunctions, but naturally occurring variations in the human genome. Autism, and in particular its mildest form Asperger's syndrome, is something that should be extolled as a cognitive variant that privileges a subset of the general population with exceptional memory skills, heightened sensory perception, and an enhanced understanding of systems. They argue that autism has been part of human evolution since the prehistoric era and credit it with allowing humanity itself to develop the means with which to cope with difficult conditions throughout time. In essence, what is being achieved is a narrativization of autism that is completely divorced from pathology and individual suffering.

On the forefront of this retelling of the autism narrative is Professor Tony Attwood who 'believes the "out of the box" thought processes of people on the autism spectrum will solve the world's big problems. He is credited with being the first clinical psychologist to present Asperger's syndrome not as something to be "fixed " but as a gift, evidenced in many of the great inventors and artists throughout history' (Australian Story). Steven Silberman, another passionate advocate for those with autism, identifies those who exhibit its characteristics as technically brilliant individuals whose social eccentricity marks them out as the ultimate non-conformists. Rather than being a social deficit, autism grants them membership into an exclusive 'neurotribe'.

Autism dates back to 1943 to a thesis produced by Professor Hans Asperger for the University of Vienna. Asperger's study was conducted at the height of the Nazi eugenics programme and the group of children he analysed were then being scrutinised for compulsory euthanasia. Asperger coined the term *Autistische Psychopathen* (autistic psychopathy) or *Autismus* to describe what he observed of these children's unusual behaviour. Often these observations bordered on the fantastic. Asperger observed 'children with the minds of geniuses, eccentrics, obsessed with their special interests, some with amazing memories who could recall all the routes of the Viennese tramlines, others who could perform rapid arithmetical calculation, and others with profound learning difficulties' (Baron-Cohen).

It is purported that Asperger saw his research subjects 'as potential innovators, seeing the world with a fresh perspective, and called them his "little professors". He suggested to his superiors that his 'little professors' would make superior code breakers for the Reich' (Baron-Cohen). Silberman concludes that 'Asperger's prediction that the little professors in his clinic could one day aid in the war effort had been prescient, but it was the Allies that reaped the benefits' (243). Little account here is taken of the eugenic context of Asperger's study, nor its premise in assuming his subjects suffering from a pathology that would often result in the societal exclusion in one form or another. Indeed, it was their failure to socially integrate and take part in community that lead the scientists working alongside Asperger to elect them in the first place as candidates for annihilation and indeed many children outside of his study had surely already been marked for death. What Asperger is suggesting is that this neurodivergent population could be exploited for military advantage and therefore has biopolitical value to the Third Reich. It is on these grounds that he recommends they should be granted a stay of execution.

Steve Silberman's assertion of a future application of these children's capabilities simply does not add up chronologically, unless this is the intention of Asperger's remarks, because of his reference to the Allies and their employment of Alan Turing. Herwig Czech paints a very different portrait of Asperger, not as a social researcher 'concerned with helping them live up to their full potential' as is described by Silberman, but rather as a diagnostician wholly complicit with the National Socialist project of race hygiene (Silberman, 129). The milieu in which Asperger operated cannot be separated from its aims. Czech's critical examination of Asperger's life concludes that he 'actively cooperated with the child "euthanasia" program. The language he employed to diagnose his patients was often remarkably harsh (even in comparison with assessments written by the staff at Vienna's notorious Spiegelgrund "euthanasia" institution), belying the notion that he tried to protect the children under his care by embellishing their diagnoses'.

What Asperger was adhering to was not a sympathy for his patients but rather a clinician scale that measured their viability within a spectrum of intellectual and other capabilities. This was totally in accord with the mission within the Nazi state, with 'its emphasis on turning troubled children into useful members of the German body politic' (Czech). A logic of social worth informs the diagnosis of Asperger's from its inception dealt with children who were considered abnormal and privileged them if they were able to exhibit normal or even above-average intellectual abilities. Those with potential could be put to work, while those whose disabilities were classed as severe were put to death. Therefore, the concept of the autism spectrum itself may be identified as contributory to genocide.

The logic of the contemporary neurodiversity movement perpetuates a judgement of what constitutes 'normality' founded upon scientific measures of health and ability. By describing high functioning autism as a desirable, inheritable trait its logic intersects with the eugenics at play in Asperger's diagnostic premise, wherein those judged to be the fittest were granted the further attributes of genius and utility. By lobbying for autism's categorical inclusion into the spectrum of normativity it remains in collusion with a regime of scientific regulation currently directed through the disciplines of psychiatry and genomics. These entities are now themselves colluding with neoliberal economics to deregulate the diagnosis of autism, making of those on the spectrum a new market constituency who can now tailor their learning, labour, and lifestyle to complement their abilities out of the remains of what was formerly considered a mental affliction. Anne McGuire argues that in recent years, neoliberalism 'has been actively blurring of categorical boundaries separating normalcy and disorder' as a means 'of stimulating the market, while also grounding the emergence of novel subjectivities and forms of normative surveillance and control' (403). The inclusion of the autism spectrum into a broader spectrum of what constitutes normativity suggests that the market has now found autism 'useful', insofar as it relates to the greater management of body, mind, and movement in the twenty-first century.

Unlike the economic environment of Nazi Germany, where 'due to increasing labor shortages, it became a political and military imperative to rehabilitate as many potential workers as possible, even those considered of inferior hereditary quality,' in the neoliberal context, rehabilitation is about judging an individual's potential for self-betterment as it relates to their potential use as a social and material resource (Czech). The severity scale measuring a level of support needed for those on the autism spectrum, therefore, in a neoliberal context, relates to a concept of biopolitics insofar as those unable to prove their capacity for independence, face a consequential threat to their continued maintenance and even existence. Those who require life-sustaining assistance are in the most precarious position. As a consequence, things like welfare, housing, social, services, education, medical care, citizenship, and social inclusion are no longer offered to them by a neoliberalised state apparatus whose aim is now fixed on eliminating access to such resources. This is the case, McQuire explains because 'the neoliberal spectrum has little use for lost causes' (417). Therefore, debility only has value if it can prove that it is profitable to capitalism and as such what is sold to those who are on the spectrum are the means through which to recover from it, overcome it, or compensate for it.

McQuire argues that this is of significance because autism as a spectrum disorder has 'long tentacles' and 'sticky labels' that are poised to relate to a 'systematic pathologization of everyday life' and with it attendant forms of oppression (412). These are seldom recognised within a neoliberal discourse of inclusion, optimism, improvement, recovery, and resiliency which together form a highly lucrative narrative around making neurodiversity positive. Those on the spectrum are therefore measured according to not just to the severity of their disorder, but rather by their capacity to cope with it, in order to strive towards a life ultimately of greater normativity. Their ability to work flexibly and to accumulate capital, as well as recognition in the world, speaks to a society orientated towards the prospect of upward mobility. Herein lies the first steps towards a relentless project of self-care and self-improvement, in which through the incremental bettering of habits and behaviour, anyone, including those who are on the spectrum, can achieve their potential.

The entrepreneurial self is by definition one plagued from the beginning with malady. These limitations, however, can be overcome through a combination of therapeutic remediation and technological enhancement that when combined with self-surveillance, productivity and consumption prepares the ground for social mobility (McQuire 418). Thus, bodies find themselves on the whole reinvigorated by neoliberal capitalism, and able to progress on a spectrum towards progress and self-betterment through their concentrated efforts. There are, of course, losers as well as winners in this system of neoliberal governance, with the biggest losers those who simply cannot get better. Those falling in that category are subject to market exposure and state censure. Again, we find ourselves in the historical territory of the management of human life and occupied by the calculation of human value against the austere arithmetic of elimination.

The question then arises do things 'get better' or do they otherwise move toward the normative requirement of complete invariably when they are influenced by systems of privilege and oppression that adapt themselves to match to the unavoidable vulnerabilities of human embodiment. In 1933, the year Hitler rose to power, Benjamin wrote of the introduction of a 'new, positive concept of barbarism' which forces individuals 'to start from scratch; to make a new start, to make a little go a long way; to begin with little and to build up further, looking neither left or right' ("Experience and Poverty" 733). He referred to these individuals as 'constructors' (733). Benjamin gives the example of Albert Einstein,

> who was not interested in anything in the whole wide world of physics except a minute discrepancy between Newton's equations and the observation of astronomy. And from this same insistence on starting from the very beginning also marks artists when they followed the example of mathematicians and built the world from

> stereometric forms, like the Cubists, or modeled
> themselves on engineers, like Klee. For like any good car,
> whose every part, even the bodywork, obeys the needs of
> the above of all of the engine, Klee's figures too seem to
> have been designed on the drawing board, and even in
> their general expression they obey the laws of the interior.
> Their interior, rather than their inwardness; and this is
> what made them barbaric (733-734).

It is that quality of interiority as opposed to inwardness, that would soon after shift the whole operating system of humanity from a fixation on inheritance to an obsession with the progress of mankind, which would inexorably lean towards the path towards annihilation. If they were alive those today those famous constructors of Benjamin's argument, Einstein and Klee, would likely be classed as neurodivergent. Neurodivergency, as a condition, bears relation to the interior on a number of levels that correspond with an appreciation of the inhuman as well as a capacity to interact with the non-human phenomena on a level equal to itself. Such equivalence brings autism into proximity with cryptography, so long as we are able to take an aptitude for codebreaking with this population seriously enough to grasp how it allows the individual to fundamentally break with an older symbolic order and in its place, install a higher functioning method of informational acquisition. Its proprietary domain name, Asperger's, is now synonymous with the non-representational, as it is the ability to correspond well with both the human in the appearance of a white, male normativity, and in the inhuman in the semblance of the robotic. The autistic in the sense represents a bridge design into the future of human evolution.

Almost a century on from Benjamin's sketch of the constructor, it is probable that such a type has abandoned drawing for programming; on his preferred surface now hangs practical information. It is also, not incidentally a male name that gets attached to this project of the interior, abandoning a former female-like characteristic of inwardness when it comes to the narration of the world. This condition rather, sides in favour of a diagnostic framework of interiority, where things can be put together and taken apart systematically, compartmentally coexisting without conflicting overlap. Those who possess alterity of sex or race, simply cannot be accommodated within such an interior, as they are seldom portrayed as persons with a divergent neurotype. Behaviour consisting of repetitive actions or movements, sound sensitivity, or diverging use of vocal tone, if detected within these subpopulations are often perceived as socially fatal glitches.

Asperger's bears within its founding lineage the markers of white supremacy. These markers can be traced back to World War II, and indeed much further back into the genealogy of race itself. 'Asperger characterized his high-functioning patients by their "finely boned features," of "almost aristocratic appearance"' (Barahona-Corrêa and Filipe qtd. in de Hooge). These characteristics associated with white superiority have carried over to 'present-day neo-Nazi circles', where 'Aspies are overtly associated with whiteness' (de Hooge). Anna De Hooge argues that that association of Asperger's with whiteness, as well as evolutionary superiority reveal 'a kinship between Aspie supremacy and Social Darwinism.' Likewise, the idea that Asperger's exists '"as the next and better phase of human evolution"' resonates with a desire to further the concept of a supremacy born in nature and born out through social recognition (Heilker qtd. in de Hooge). Evolutionary hierarchy, in this sense, remains not only raced but pre-programmed to conform to existing standards where statistics bear out the fact that Asperger's 'the sex ratio is at least ten males to every female' (Baron-Cohen quoted in de Hooge). Similarly, statistics bear out the fact that Asperger's is the highest amongst whites, far outpacing the diagnoses recorded in Latino, black and Asian groups. It is possible, however, that those diagnosing Asperger's already carry a diagnostic bias with them that suggests that their proto-typical subject is male, economically privileged and white. Such assumptions do much to reify in themselves, the concept that white male subjects will continue to occupy a functionally higher status than the female subject, or the black subject of clinician interest. Bodies in the future will continue to matter so long as we have a latent interest in their standard issue being classified as white and their being made operational within a spectrum of white normativity.

Benjamin observed, nobody wants to drag along 'an image of man festooned with all the sacrificial offerings of the past,' but rather 'they turn instead to the naked man of the contemporary world who lies like a new born babe in the dirty diapers of the present' ("Experience and Poverty" 733). As a consequence, the future must be about changing reality instead of describing it and time its evolutionary guise made functional to this cause of historical revisionism. Within this scenario, it is possible to imagine the (white) man of history as one burdened by the guilt of his various acts of barbarity, as well as his desire in the modern age to throw off those various sins of the past and start again bearing a façade of innocence. In order for this to happen, he must be ostensibly willing to allow others to mount the historical stage, decentring himself from a pride of place as the sole agent shaping the course of reality and take a renewed interest in the contributions of others. In the present neoliberalism functions to decentre subjects this way, making them essentially relational to one another.

This situation has particular resonance when it comes to the valuation of an 'autistic' sensibility within the context of a mankind that is now actively evolving to conform to the intelligence standard of information. Steven Shaviro argues that what elevates those on the autistic spectrum from their ordinary counterparts is their ability to view the world around them with empathy and concern, and that indeed, 'autistics are in fact acutely sensitive *beyond* the human' ("Thinking Blind" 323 my emphasis). Their advanced sensory capability allows them to be 'responsive to "resonances across scales and registers of life, both organic and inorganic ... and 'the testimony of autistics themselves indicates that, for them, "*everything* is somewhat alive"' (323). Through such comments, Shaviro indicates that those on the spectrum might be capable of relating to artificial forms of intelligence in a way that is superior to other humans and moreover that they uniquely can appreciate the presence of a multiplicity of intelligence hovering in their midst.

Ordinary humans, by contrast, lack this form of super sentience, if you like, because they continue to separate that act of thinking from being, and in doing so, maintain a perception of reality based solely through the correlation of these two. This does not allow the possibility of reality to exist beyond the human mind. The autistic mind, however, is able to overcome the anthropocentric bias of this assumption 'that the world exists essentially or exclusively *for us*' and thus is able to extend the concept of reality making to include nonhuman others (323). Rather than being deficient in sensibility, autistic individuals are uniquely alive to the possibility that there is a reality behind human thought and as such objects might relate to one another in various ways that are beyond conventional human perception. Moreover, they are able to perceive that objects can possess qualities that exist independently from the human senses. We might reference autistics then as a class of philosophical early adopters able to cope with better than most the physical and technological implications of the Anthropocene and a new 'ecology that does not privilege the human but attends to the more-than-human"' (324). This situation implies that mankind is now, in fact, looking to engage with forms of life potentially superior to itself. This line of thought is now commonly referred to as 'speculative realism'. Speculative realists like Shaviro, insist that the 'world in itself the world as it exists apart from us cannot in any way be contained or constrained by the question of our access to it' ("Panpsychism"). Nevertheless, access, at whatever level, remains of key significance as a tool of exchange within the context of neoliberal realism.

During a recent speech that teenage climate change activist Greta Thunberg gave to MPs at the Houses of Parliament, she repeatedly asked, 'is my microphone on? Can you hear me?', 'Did you hear what I just said? Is my English OK? Is the microphone on? Because I'm beginning to wonder' and finally, 'I hope my microphone was on. I hope you could all hear me.' This affect could be said to be related to Thunberg's Asperger's. That said perhaps it was not in the way that one would conventionally imagine. '"Being different is a gift," she told Nick Robinson when interviewed on Radio 4's Today programme. "It makes me see things from outside the box. I don't easily fall for lies, I can see through things. If I would've been like everyone else, I wouldn't have started this school strike for instance" '(Thunberg qtd in Birrell). What Thunberg is suggesting is that she is not limited to human judgement, nor centred on human subjectivity in particular when she embarks upon her activism. Rather, she is able to achieve a form of sentience around climate change that few others are able to apprehend. For Thunberg, there is a fundamental lack of distinction between the milieu in which is speaking and the greater organism of the planet to which it must crucially relate.

Thunberg's parents relay the fact that as a child she suffered from profound shyness and depression (Birrell). Her aptitude for public speaking is a newly acquired trait, whose acquisition relates directly to her 'remorseless focus on the core issue of climate change' (Birrell). Therefore, she passionately believed someone had to speak on not on behalf of nature as well as generation, but *as* nature and generation. If we revisit Thunberg's questions, it is possible to admit that she is addressing a constituency beyond her immediate audience, who indeed break into nervous affirmative laughter at the very mention of her anxiety at being not only heard but understood. She wonders if the parties she is ostensibly addressing have the capacity to understand her, rather than about their capacity to literally hear her. One can speculate that Thunberg has, indeed, other audiences on her mind beyond this one to which she strives to achieve transmission of her ideas. Some of these might not be human, or they might more accurately speaking be depersonalised beings that like her have the ability to sense things differently beyond what is considered by human standards a typical capacity of attention.

What Thunberg is doing is speaking the language of the algorithm and carefully curating the data points of her life story. This has very little to do with audiences such as those she encountered here. Thunberg is addressing a far greater online constituency in crafting her message on climate change to coincide at every point with the steps taken to achieve dominance of social media stratification. Working alongside her mother, the Swedish opera singer and celebrity Malena Ernman they have produced a compelling image of Thunberg as kind of Asperger's superhero whose unique power is her ability to emotionally access to the suffering of the natural world. They provide a genetic basis to their argument by pointing out that her father the actor Svante Thunberg counts the Nobel Prize winner, Svante Arrhenius amongst his esteemed ancestors. Arrhenius was a Swedish physicist and chemist who received the Nobel Prize for Chemistry in 1903. Ironically, the Nobel Foundation, in the same year that Thunberg started her movement, 2016, became the subject of a campaign by Fossil Free Sweden to compel it to divest from investment in fossil fuels ("Nobel Prize").

Thunberg's origin story begins at the site of her own physical starvation and mutism before she is clinically saved through her accurate diagnosis. Within this realisation came her calling; her ability to publicly link the narrative of personal crisis to that of planetary crisis. Thunberg frames her personal crises in a way that posits her suffering as equivalent to 'the oppression of women, minorities, and people with disabilities', who experience similar levels of stress that all conveniently arc back to global climate crisis, and therefore their maladies are all classed as 'symptoms of the same systemic disorder' (Neuding). What this narrative also makes evident is that we aren't really to care about these other people's stories, with intimate detail applied to their own suffering, but for everyone regardless of who they are to go out and 'change the system' so that presumably Thunberg and the others can finally get better. What is missing from this narrative is any serious challenge posed to the economics of attention that is the real currency of our information age. Or the fact that Thunberg's narrative of suffering corresponds perfectly with a neoliberal affective economy that privileges those who are already economically and socially advantaged to aim higher still for recognition of their likes and dislikes, their feelings within a digital world that progressively banks these.

The Financial Climate

Thunberg's Extinction Rebellion movement is closely aligned to the for-profit climate change consultancy, *We Don't Have Time*. '*We Don't Have Time* is mainly active in three markets: social media, digital advertising and carbon offsets' (Morningstar). Within the trifecta of sustainability rebranding for the twenty-first century, Thunberg stands out as its chief generational influencer. Identified as a disruptive startup, *We Don't Have Time* uses social networking as a way to stimulate a new generation of youth to invest in the principle of carbon off-setting that is guaranteed in its efficacy through the platform's own certification. Individuals, organisations and corporations are all invited to join in this public activism via private consumption.

Their model appeals to the concept of a grassroots communication, that if successfully redeployed in this manner will ultimately yield billions in wealth volunteered by Thunberg's young followers to finance the growth and development of the new green economy by purchasing high climate rated products as verified on the *We Don't Have Time's* own platform and validated by its membership. They can esteem this contribution because they can feel confident that through their purchase they are acting as their own saviours. In reality, they are acting in the role that governments used to by providing massive public funding to private industry. Therefore, while Thunberg espouses she is putting an emergency brake on capitalism through her activism, in reality, she is propelling it forward by using a different fuel source that is equally pernicious.

We Don't Have Time's ultimate business plan is not to save the planet, but to save capitalism through the monetisation of natural processes i.e. carbon emissions, and the trade of them in the global market. This economy will be complemented by the introduction of a host of new environmental remediation technologies that will be sold to poorer nations who not coincidentally will bear the brunt of climate change. Once again, with the global North acting as the global South's creditor. Colonialism, this time around will be draped in green. These nations have no equalivent Thunberg at the helm. It is not coincidental that Sweden the nation of Thunberg's birth, has pledged to go toward 100% renewable energy by 2040, and be the first amongst the European Nordic countries to do so. Norway, Denmark, Germany, and Finland follow closely behind according to the aptly named 'bi-annual EY Renewable energy country attractiveness index (RECAI)' (Sims). Thunberg then becomes very much the attractive face of a second nature founded on renewables.

We Don't Have Time is a social engineering project that employs Thunberg's image to draw attention to a Eurocentric humanity in crisis that is now in many ways intent on preserving the comforts that industrial capitalism historically afforded it. The lack thereof is profoundly threatening that population and forcing it to act to save itself through what it perceives as a radical means. Such means are achieved through the wholesale redefinition of nature itself to conform to the standards of technology. It is therefore not inconceivable that the generation following on from Thunberg's will come to associate nature with 'wind turbines and solar panels', rather than 'trees and insects' (Davey). The consequence of which will be a wholesale rebranding of nature, wherein these commercial products assume a symmetry with the image of an outside world. Buckminster Fuller's geodesic domes of the 1970s very much foreshadowed the coming of this technology as a necessity to preserve the functioning of our "Spaceship Earth". The fuel which will keep us aloft is none other than capital retooled for the information age.

A geodesic dome literally features in a *We Don't Have Time's* promotional YouTube video. Each joint features the carefully distributed racialised emoji faces of young people that are shown to make up a social network; the geodesic dome figuratively becoming the architecture in which such a development is based (*"We Don't Have Time"*). Similar to the vague inclusion of those faces, *We Don't Have Time* is only superficially in the market of climate activism. Rather like its Silicon Valley cultural counterparts, its deeper commercial ambitions lie in a domination of social media and digital advertising. *We Don't Have Time* is built on the same principles as those used by the likes of Facebook and Twitter, to promote the expression of personal awareness and social activism as a means of generating saleable data, as well as, the promotion of related products and services.

When Thunberg implores world leaders in Davos to feel 'the panic and fear she feels everyday' it is worth noting how much these emotions are used to reify certain natures (Thunberg). Indeed, through a rating system devised by *We Don't Have Time* users are encouraged to either love or hate those in political power and indeed, to in their own words 'climate bomb' those that fail to act to mitigate carbon levels. What these figures are bombarded with is the element of carbon itself. In other words, destroyed by the very thing they failed to eliminate as 'climate villains'. Their shortcomings will be exposed through social media, and as a consequence, their careers will be blown up as retribution for their failure to act (*"We Don't Have Time"*).

Users of the platform are encouraged act like Thunberg, as a superhero on an urgent mission to take down the bad guys. What is perhaps most curious about this emergency vigilante mode We Haven't Got Time proposes for the social media public to enter is how closely it resembles some of the behavioural traits of autism. In the words of Thunberg herself which she reiterates to her followers, 'our senses are heightened, you are focused like a laser, and you put your entire self into your actions' (Salamon). This quote comes from Margaret Klein Salamon, the Founder and Director of *The Climate Mobilization*, who describes herself as a climate psychologist. Her Foundation has now publicly aligned with Thunberg's Extinction Rebellion Movement. In her own work, she compares climate crisis to America's total mobilisation of its civilian population during World II. Klein Salamon points to several financial phenomena that took place at that time which her mobilisation efforts aim to replicate:

> Citizens invested their available cash reserves in war bonds. Taxes were also increased significantly, particularly on high earners, who paid a steep "Victory Tax," the most progressive tax in American history. The top marginal income tax rate on the highest earners reached 88% in 1942 and a record 94% in 1944. A tax on excess corporate profits provided about 25% of revenues during the war. The federal government instituted a sweeping rationing program in order to ensure a fair distribution of scarce resources on the home front – and to share the sacrifice equitably. Gasoline, coffee, butter, tires, fuel oil, shoes, meat, cheese, and sugar were rationed, and every American received a fair share. "Pleasure driving" was banned, the Indy 500 was shut down, and a national speed limit of 35 miles per hour was established. Comprehensive wage and price controls were put in place to combat inflation.

Citizens invested their available cash reserves in war bonds. Taxes were also increased significantly, particularly on high earners, who paid a steep "Victory Tax," the most progressive tax in American history. The top marginal income tax rate on the highest earners reached 88% in 1942 and a record 94% in 1944. A tax on excess corporate profits provided about 25% of revenues during the war. The federal government instituted a sweeping rationing programme in order to ensure a fair distribution of scarce resources on the home front – and to share the sacrifice equitably. Gasoline, coffee, butter, tires, fuel oil, shoes, meat, cheese, and sugar were rationed, and every American received a fair share. "Pleasure driving" was banned, the Indy 500 was shut down, and a national speed limit of 35 miles per hour was established. Comprehensive wage and price controls were put in place to combat inflation.

What this list concludes is that what will be mobilised here are various instruments that further the project of the financialisation of the climate crisis. The list economic resources made available by civilians in World War II fails to make mention of one of its most significant beneficiaries: The Manhattan Project. America "ended the war," by dropping atomic bombs on the cities of Hiroshima and Nagasaki. This was an act of unprecedented human and environmental destruction. It ushered in what many scientists believe to be the dawn of the Anthropocene from the fall out of that radical military action. Indeed, it is the initial exploitation of fossil fuel energy and later atomic energy that ultimately wins the war for America. What is sacrificed by the American people is the civilian use of energy for the greater empowerment of their military. Klein Salamon's veneration of America's wartime economy does nothing to challenge its extreme energy use, nor does it criticise the structures of the military-industrial complex itself.

This trend continued from the time of the war onwards that the military continues to dominate the consumption of fossil fuels to this day, wherein 'the U.S. Department of Defense is the largest consumer of oil in the U.S. and the largest institutional consumer of oil in the world. Every year, our armed forces consume more than 100 million barrels of oil to power ships, vehicles, aircraft, and ground operations—enough for over 4 million trips around the Earth, assuming 25 mpg' (The U.S. Military and Oil"). The concept of each person reducing their 'environmental footprint', becomes laughable when we consider the issue from this dimension. However, what it does is create a certainty around the concept of responsibilisation that is extremely effective in getting people to spend their way out of crisis, not through excessive purchase of fossil fuel-based goods like cars, clothing, fast food, and the like, but rather through buy in to funds that promote a furthering of disciplinary society under the banner of climate change. This is achieved within a social media environment that is little more than an informatics of control nudging forth acceptance of various forms of conformist behavioural change.

That autism becomes the powerhouse of this model of progress suggests that things should be seen in black and white, and that the bad products, actors, and concepts that are causing climate change can be readily identified and eliminated and that those with the greatest capacity for goodwill naturally rise to the top of the rating charts. Perhaps it is only a troubled adolescent that could see things in this way, but of course at another level this entire discourse is infantilising the public to which it is directed, distracting them from the fact that they are as little in control of their fate as the average middle-class teenager. Today a minority are housed within a nuclear family unit. Rather it is within the algorithmic estate that the majority now dwell. It is this private domain that dictates the parameters of their environment in exchange for certain protections. The ends of the future have us all dwelling in a place occupied with technology as its master, enticing us to express ourselves, while at the same time carefully placating our desires. We are no longer in the era where codebreakers are needed for the security of the West. Today it is codemakers who perform that function.

The Alternative für Deutschland party (AfD), Germany's far-right party, has come out against Greta Thunberg, ridiculing her 'as "mentally challenged" and a fraud' (Connolly). The party has recently forged ties with prominent conservative groups in the US in order to undermine her growing influence beyond Sweden. AfD 'candidates have made comparisons between the Swedish teenager and a member of a Nazi youth organisation and called for her to seek treatment for what Maximilian Krah, an AfD candidate for the EU elections, called her "psychosis"' (Connolly). Most of this vitriol is now being disseminated via social media, each side fighting to become the arbiter of truth when it comes to climate change. The truth, however, has become something that is now revised daily in the form of posts and tweets and in this febrile social atmosphere little can be said to justify the fanaticism that has now come to dominate as a feature of politicised discourse. What is lost in any understanding of whom manages the spectrum itself and to what degree they are pushing the limits of these debates in directions that will ultimately constitute our world to come where social marginalisation becomes the stuff of asset and is made to function as capital. In this way, adaptation becomes the privileged mode of actualisation such that increasingly intense relations with all manner of technological objects and processes become the source of both reality making and storytelling. The natural and the artificial, the organic and the machinic, have never been as closely related and this is an effect of our becoming synonymous with our technologies, rather than being determined by them as in previous centuries.

Construed through the lens of that order, the concept of speculative realism, with its veneration of an autistic sensibility, does little to seriously challenge existing structures of power and, in particular, the way that knowledge is made operational through routine disregard of the sensibilities of women and people of colour, essentially through their objectification. The failure of the speculative realist project then in some senses is a failure to discriminate insofar as its tendency is to '"attend to everything the same way with no discrimination"' (325). Discrimination of this sort requires an ability to subtract difference and account for multiplicity. Perhaps humanity's desire to completely immerse itself in data 'bespeaks a similar inability to subtract, to simplify, and to hierarchize' in ways that do not privilege a solitary subject, revert back to an auto-referential self, or engage in normative distinctions as a means of understanding the world. The neoliberal world order in which digitisation operates, however, continues to assume that corresponds to human parameters. Therefore, data assumes universality according to a flat ontology, that remains blind to status or category even as its main imperatives remain prefiguration and classification. It has awareness, but only insofar as it is programmed to resist alterity and institute likeness. Within this scenario, a nuanced inclusion of object awareness on the part of humanity seems still a long way off. Presently, we find ourselves back in the territory of training the mind through a reorganisation of its connections. Through the concept of neural plasticity, it is becoming increasingly possible to have education resemble software development.

In this way, digital technologies can effectively repurpose formative thinking to emphasise human capital development, by coding the brain itself to conform with the aim of personal optimisation.

Autism as a sensibility exists to alert us to the fact humanity now routinely coexists and coevolves with the nonhuman through various modes of technology, and in so doing assumes capacities and potentials 'beyond the human' as a matter of course. The mind now performatively maps the affective conditions of the Anthropocene, mimicking the spaces and times under its purview and making of them too a quality unlike what has passed. The ends of the future, finally preoccupy themselves not by simply recognising a pre-existing nature, nor revealing a pre-existing world, but rather with the ways in which relations are constituted, and encounters produced. In the end, this world in its entirety becomes in its own way utterly alien to us. This will usher in a new era of deep time, which invariably forges a pretext for colonisation when humanity inevitably fail to be adequate custodians of this world and must seek the fundamental neutrality of another. Except for the fact that nothing is really over and the Earth's envelope only possesses the illusion of closure. Ours will have to become a productive pathology to cope with the uncertainty tied to such an understanding. Acceding to the demands induced by the loss of experience and increase of exploitation thrusts us into a future where, in the end, all of existence must become operative.

Works Cited

Adams, Stephen B. "Arc of Empire: The Federal Telegraph Company, the US Navy, and the Beginnings of Silicon Valley." *Business History Review,* vol. 91, no. 2, 2017. pp. 329-359.

https://doi.org/10.1017/S0007680517000630. Accessed 12 Nov 2018.

"A Day in the Life The Beatles." *Genius,* n.d. https://genius.com/The-beatles-a-day-in-the-life-lyrics. Accessed 15 Jun 2019.

Adeleke, Tunde. *UnAfrican Americans: Nineteenth-Century Black Nationalists and the Civilizing Mission.* University Press of Kentucky, 2015.

Agamben, Giorgio. "For a Theory of Destituent Power." Public lecture in Athens. 16 Nov 2013. Transcript of lecture delivered by Giorgio Agamben in Athens, 16 Nov 2013. *Critical Legal Thinking,* 5 Feb 2014, http://criticallegalthinking.com/2014/02/05/theory-destituent-power/. Accessed 18 July 2017.

Ahmed, Nafeez. "No Piers Morgan – This is how to destroy the Islamic State." *Middle East Eye,* https://www.middleeasteye.net/opinion/no-piers-morgan-how-destroy-islamic-state5 Feb 2015. Accessed 24 Dec 2015.

Alexander, Elizabeth. "'Can you be Black and Look at This?': Reading the Rodney King Video(s)." *Public Culture,* vol. 7, no. 1, 1994, pp. 77-94. https://doi.org/10.1215/08992363-7-1-77. 2 Jun 2015.

Amin, Samir. *Eurocentrism: Modernity, Religion, and Democracy.* Monthly Review Press, 2009.

Armiero, Marco. "Or, In Praise of Mutiny." *Future Remains: A Cabinet of Curiosities for the Anthropocene.* Edited by Gregg Mitman, Marco Armiero, Robert Emmett, University of Chicago Press, 2018, pp. 129-140.

Armitage, David. "The Elizabethan Idea of Empire." *Transactions of the Royal Historical Society,* vol. 14, 2004, pp. 269–277. *JSTOR,* www.jstor.org/stable/3679320. Accessed 4 Feb 2015.

Armitage, John and Phil Graham. "Dromoeconomics: Towards a Political Economy of Speed." *Parallax,* vol. 7, no. 1, 2001, pp. 111-123, https://doi.org/10.1080/13534640010015999. Accessed 16 Dec 2015.

Atanasoski, Neda, and Kalindi Vora. "Surrogate Humanity: Posthuman Networks and the (Racialized) Obsolescence of Labor." Catalyst: Feminism, Theory, Technoscience, vol. 1, no. 1, 2015, https://doi.org/10.28968/cftt.v1i1.28809. Accessed 20 Dec 2015.

Attali, Jacques. *Noise: The Political Economy of Music.* Translated by Brian Massumi. University of Minnesota Press, 1985.

Atwan, Abdel Bari. Islamic State: *The Digital Caliphate.* University of California Press. 2015.

Baldwin, Robert C.D. "Colonial Cartography under the Tudor and Early Stuart Monarchies, ca. 1480–ca. 1640." *The History of Cartography, Volume III, Cartography in the European Renaissance.* Edited by David Woodard, University of Chicago Press, 2007, pp. 1754-1779.

Ball, Philip. "The maddeningly magical maths of John Dee." *New Scientist,* 24 Feb 2016, https://www.newscientist.com/article/2078295-the-maddenin gly-magical-maths-of-john-dee/. Accessed 10 Aug 2017.

Baron-Cohen, Simon. "Did Hans Asperger save children from the Nazis — or sell them out?" *The Spectator.* 12 Sept 2015, https://www.spectator.co.uk/ 2015/09/did-hans-asperger-save-children-from-the-nazis-or-sell-them-out/. Accessed 10 May 2019.

Barrio-Vilar, Laura. "'All O' We Is One'?: Migration, Citizenship, and Black Nativism in the Postcolonial Era." *Callaloo,* vol.37, no. 1 Dec 2014, pp. 89-111., doi: 10.1353/cal.2014.0019. Accessed 5 May 2016.

Beidler, Philip. "China Magic: America's Great Reality Hiatus, 1948-73."*Michigan Quarterly Review,* vol. 47, no. 2, 2008, pp. 150-0_7. *ProQuest,* http://proxy.cca.edu/login?url=https://search-proquest-com. proxy.cca.edu/docview/232312989?accountid=30962. Accessed 29 Jan 2019.

Benjamin, Walter. "Surrealism." *Walter Benjamin Selected Writings, 1927-1930.* Edited by Rodney Livingstone, Michael William Jennings, Howard Eiland, and Gary Smith Harvard University Press, 2005, pp. 207-221.

--- . "The Destructive Character." *Walter Benjamin: Selected Writings. Vol. 2 Part 2, 1931-1934.* Edited by Michael William Jennings, Howard Eiland, and Gary Smith, Translated by Rodney Livingstone and Others, Belknap of Harvard UP, 1999, pp. 541-542.

---. "Mickey Mouse." *Walter Benjamin: Selected Writings. Vol. 2 Part 2, 1931-1934.* Edited by Michael William Jennings, Howard Eiland, and Gary Smith, Translated by Rodney Livingstone and Others, Belknap of Harvard UP, 1999, pp. 545-546.

---. "Berlin Chronicle" *Walter Benjamin: Selected Writings. Vol. 2 Part 2, 1931-1934.* Edited by Michael William Jennings, Howard Eiland, and Gary Smith, Translated by Rodney Livingstone and Others, Belknap of Harvard UP, 1999, pp. 595-635.

---. "Hashish in Marseilles." *Walter Benjamin: Selected Writings. Vol. 2 Part 2, 1931-1934.* Edited by Michael William Jennings, Howard Eiland, and Gary Smith, Translated by Rodney Livingstone and Others, Belknap of Harvard UP, 1999, pp. 673-679.

---. "Experience and Poverty." *Walter Benjamin: Selected Writings. Vol 2 Part 2, 1931-1934.* Edited by Michael William Jennings, Howard Eiland, and Gary Smith, Translated by Rodney Livingstone and Others, Belknap of Harvard UP Press, 1999, pp. 731-738.

---. "Hitler's Diminished Masculinity." *Walter Benjamin: Selected Writings. Vol 2 Part 2, 1931-1934.* Edited by Michael William Jennings, Howard Eiland, and Gary Smith, Belknap of Harvard UP, 1999, pp. 792-793.

---. "The Work of Art in the Age of its Technological Reproducibility: Second Version." *Walter Benjamin: Selected Writings. Vol. 3, 1935-1938.* Edited Michael W. Jennings and Howard Eiland, Belknap of Harvard UP, 2002, pp. 100-131.

--. "The Formula in Which the Dialectical Structure of Film Finds Expression." *Walter Benjamin: Selected Writings. Vol. 4, 1938-1940.* Edited by Howard Eiland and Michael W. Jennings, Translated Edmund Jephcott and Others, Belknap of Harvard UP, 2003, pp. 94-95.

---. "The Work of Art in the Age of its Technological Reproducibility: Third Version." *Walter Benjamin: Selected Writings. Vol. 4, 1938-1940.* Edited by Howard Eiland and Michael W. Jennings, Translated by Edmund Jephcott and Others, Belknap of Harvard UP, 2003, pp. 251-283.

---. "On the Concept of History." *Walter Benjamin: Selected Writings. Vol. 4, 1938-1940.* Edited by Howard Eiland and Michael W. Jennings, Translated by Edmund Jephcott and Others, Belknap of Harvard UP, 2003. 389-400.

---. "Paralipomena to 'On the Concept of History.'" *Walter Benjamin: Selected Writings. Vol. 4, 1938-1940.* Edited by Howard Eiland and Michael W. Jennings, Translated Edmund Jephcott and Others, Belknap of Harvard UP, 2003, pp. 401-411.

---. *Illuminations.* Houghton Mifflin Harcourt, 1968.

---. *Reflections: Essays, Aphorisms, Autobiographical Writings.* Mariner Books, 2019.

---. *The Correspondence of Walter Benjamin, 1910-1940.* University of Chicago Press, 1994.

---. *The Origin of German Tragic Drama.* Verso, 2003.

Bengal, Rebecca. "The Interstellar Style of Sun Ra." *The Pitchfork Review*, 18 Apr 2016. https://pitchfork.com/features/from-the-pitchfork-review/9866-the-interstellar-style-of-sun-ra/ Accessed 17 Jun 2019.

Berardi, Franco. *After the Future.* AK Press, 2011.

Berger, J.M. "The Metronome of Apocalyptic Time: Social Media as Carrier Wave for Millenarian Contagion." *Perspectives on Terrorism*, vol. 9, no. 4, 2015, pp. 61–71. *JSTOR*, www.jstor.org/stable/26297415. Accessed 24 Dec 2015.

Berry, David M., and Alexander R. Galloway. "A Network Is a Network Is a Network: Reflections on the Computational and the Societies of Control." *Theory, Culture & Society*, vol. 33, no. 4, July 2016, pp. 151–172, doi:10.1177/0263276415590237. Accessed 20 Dec 2015.

Birrell, Ian. "Greta Thunberg teaches us about autism as much as climate change." *The Guardian.* 23 Apr 2019, https://www.theguardian.com/commentisfree/2019/apr/23/greta-thunberg-autism. Accessed 12 May 2019.

Black, Edwin. *IBM and the Holocaust: The Strategic Alliance between Nazi Germany and America's Most Powerful Corporation.* Crown Books, 2001.

Bratton, Benjamin H. *The Stack: On Software and Sovereignty.* The MIT Press, 2015.

Breight, Curtis C. *Surveillance, Militarism and Drama in the Elizabethan Era.* Palgrave Macmillan, 1996.

Browne, Simone. "Digital Epidermalization: Race, Identity and Biometrics." *Critical Sociology*, vol. 36, no. 1, Jan. 2010, pp. 131–150, doi:10.1177/0896920509347144. Accessed 30 May 2016.

---. *Dark Matters: On the Surveillance of Blackness.* Duke University Press, 2015.

Bruun, Ole. *Fengshui in China: Geomantic Divination between State Orthodoxy and Popular Religion.* University of Hawaii Press, 2003.

Carrington, Damian. "The Anthropocene epoch: scientists declare dawn of human-influenced age". *The Guardian.* 29 Aug 2016. https://www.theguardian.com / environment / 2016 / aug / 29 / declare - anthropocene - epoch-experts-urge-geological-congress-human-impact-earth. Accessed 1 March 2019.

Chandler, David. "Reconceptualizing International Intervention: Statebuilding, 'Organic Processes' and the Limits of Causal Knowledge." *Journal of Intervention and Statebuilding*, Issue 1, Mar 2015, pp 1-19. https://doi.org/10.1080/17502977.2015.1015247. Accessed 18 July 2017.

Chow-White, Peter A. "The informationalization of Race: Communication, Databases, and the Digital Coding of the Genome." *International Journal of Communication*, vol. 2 2008, pp, 168-1194, doi: 1932-8036/20081168. Accessed 28 May 2016.

Chude-Sokei, Louis. *The Sound of Culture: Diaspora and Black Technopoetics.* Wesleyan University Press, 2015.

Chun, Wendy Hui Kyong. "Introduction: Race and/as Technology; or, How to Do Things to Race." *Camera Obscura 70*, vol. 24, no.1, 2009, pp.7-35. https://doi.org/10.1215/02705346-2008-013. Accessed 3 March 2016.

Coates, Ta-Nehisi. *Between the World and Me.* Spiegel and Grau, 2015.

Colebrook, Claire. "Anti-Catastrophic Time." *New Formations*, 2017, pp. 102 - 119, doi: 10.3898/NEWF:92.07.2017. Accessed 28 Apr 2019.

Colebrook, Claire. "The Future of the Anthropocene." Unpublished draft. *Academia.edu*, https://www.academia.edu / 36246542 / The_Future_of_ the_Anthropocene. Accessed 8 May 2019.

Coley, Rob. "Vector Portraits, Or, Photography for the Anthropocene." Philosophy of Photography, vol. 6, no. 1-2, 2015, pp. 51-60. Accessed 27 July 2017.

Colman, Felicity J."Dromospheric Generation: The Things That We Have Learned Are No Longer Enough." *Cultural Politics*, vol. 11 no. 2, 2015, pp. 246-259. *Project MUSE*, muse.jhu.edu/article/595003.Accessed 16 Dec 2015.

Colomina, Beatriz. "Enclosed by Images: The Eameses' Multimedia Architecture." *Grey Room*, no. 2, 2001, pp. 7–29. *JSTOR*, www.jstor.org/stable/1262540.Accessed 15 Apr 2019.

Connolly, Kate. "Germany's AfD turns on Greta Thunberg as it embraces climate denial." *The Guardian.* 14 May 2019. https://www.theguardian.com/ environment/2019/may/14/germanys-afd-attacks-greta-thunberg-as-it-embraces-climate-denial. Accessed 14 May 2019.

Cormack, Lesley B. "Britannia Rules The Waves?: Images of Empire in Elizabethan England." *Early Modern Literary Studies Special Issue 3*, vol. 4, no. 2, September, 1998, pp. 10.1-20. http://purl.oclc.org/emls/04-2/cormbrit.htm. Accessed 28 Sept 2015.

Cramer, Florian. "What is 'Post-digital'?." *Postdigital Aesthetics*. Edited by David M. Berry and Michael Dieter, London: Palgrave Macmillan, 2015, pp. 12-26.

Critchley, Simon and Jamieson Webster. *Stay, Illusion!: The Hamlet Doctrine.* Pantheon Books, 2013.

Critchlow, Andrew. "Can Elon Musk and Tesla save the mining industry?" *The Telegraph.* 2 May 2015, https://www.telegraph.co.uk/finance/ newsbysector/industry/mining/11621896/Can-Elon-Musk-and-Tesla-save-the-mining-industry.html. Accessed 10 March 2019.

Curtin, P. D. "'SCIENTIFIC' RACISM AND THE BRITISH THEORY OF EMPIRE." *Journal of the Historical Society of Nigeria*, vol. 2, no. 1, 1960, pp. 40–51. *JSTOR*, www.jstor.org/stable/41970819.Accessed 15 May 2016.

Czech, Herwig. "Hans Asperger, National Socialism, and "race hygiene" in Nazi-era Vienna." *Molecular Autism*, vol.9, no. 29. 18April 2018. https://doi.org/10.1186/s13229-018-0208-6. Accessed 2 May 2018.

Darda, Joseph. "Post-Traumatic Whiteness: How Vietnam Veterans Became the Basis for a New White Identity Politics." *Los Angeles Review of Books*. https://lareviewofbooks.org/article/post-traumatic-whiteness-how-vietnam-veterans-became-the-basis-for-a-new-white-identity-politics/#! Accessed 20 Nov 2018.

Davey, Brian. "Greta Thunberg, PR And The "Climate Emergency." *Wrong Kind of Green.* 7 May 2019 http://www.wrongkindofgreen.org/2019/05/07/ greta-thunberg-pr-and-the-climate-emergency/. Accessed 10 May 2019.

Davis II, Charles. "Black Spaces Matter." Aggregate. Volume 2, March, 2015, pp. 1-8. http://www.we-aggregate.org/piece/black-spaces-matter. Accessed 13 April 2019.

Davis. Mike. *Late Victorian Holocausts: El Niño Famines and the Making of the Third World.* Verso, 2001.

Dawes, Simon. "The Digital Occupation of Gaza: An Interview with Helga Tawil-Souri." *Networking Knowledge, Special Issue: Mediatizing Gaza*, vol. 8, no. 2, 2005, pp. 1-18, doi: https://doi.org/10.31165/nk.2015.82.374. Accessed 13 May 2017.

Day, Ronald, E. *The Modern Invention of Information: Discourse, History, and Power.* Southern Illinois University Press, 2001.

De Hooge, Anna N. "Binary Boys: Autism, Aspie Supremacy and Post/Humanist Normativity." *Disability Studies Quarterly.* vol. 39. no.1, 2019, http://dsq-sds.org/article/view/6461/5185. Accessed 4 May 2019.

Deese, Richard S. "The artifact of nature: 'Spaceship Earth' and the dawn of global environmentalism." *Endeavour.* Vol 33, no.2, June 2009, pp. 70-75. doi: 10.1016/j.endeavour.2009.05.002. Accessed 5 May 2019.

Deleuze, Gilles and Felix Guattari. *A Thousand Plateaus.* London: Continuum 1987.

Deleuze, Gilles. *The Fold: Leibniz and the Baroque.* University of Minnesota Press, 1993.

Diamandis, Peter H. "The Brain Tech to Merge Humans and AI Is Already Being Developed." The *Singularity Hub.* 5 Dec 2016, https://singularityhub.com/2016/12/05/the-brain-tech-to-merge-humans -and-ai-is-already - being - developed/#sm.000066usxhsfze7btrm2njado xoki. Accessed 15 April 2019.

Doroudi, Sherwin. "On Leibniz and the I Ching." *California Institute of Technology-Information Technology Services.* vol. 26 2007, http://ramix .org/china/LeibChi.pdf. Accessed 11 Oct 2015.

Dowd, Katie. "How the Beach Boys ended up recording a song written by Charles Manson."*SFGATE*, 20 Nov 2017, https://www.sfgate.com/bay area / article / beach-boys-song-written-by-charles-manson-12371418. php. Accessed 3 Jun 2019.

Drescher, Seymour. "The Ending of the Slave Trade and the Evolution of European Scientific Racism". *Social Science History.* vol. 14, no 3, 1990, pp. 415-45, https://doi.org/10.1017/S0145553200020861.

Accessed 15 May 2016.

Duffield, Mark. "Is the Earth Curved or Flat?" *Global Insecurities Centre, University of Bristol*. April 2017, https://www.bristol.ac.uk/media-library/sites/global-insecurities/documents/Duffield_Curved%20or%20 Flat_April%2017.pdf. Accessed 22 Sep 2017.

Eiland, Howard. "Reception in Distraction." *boundary 2*, vol. 30 no. 1, 2003, pp. 51-66. *Project MUSE*, muse.jhu.edu/article/41340. Accessed 2 Nov 2018.

Eshun, Kodwo. *More Brilliant than the Sun: Adventures in Sonic Fiction*. Verso Books, 1999.

Ezcurra, Mara Polgovsky. "On 'Shock:' The Artistic Imagination of Benjamin and Brecht." *Contemporary* Aesthetics, vol. 10, 2012, http://www.contempaesthetics.org/newvolume/pages/article.php?articleID=659. Accessed 8 Nov 2018.

Findlay, John. M. *Magic Lands: Western Cityscapes and American Culture After 1940*. University of California Press, 1993.

Fitter, Chris. "Review of Surveillance, Militarism and Drama in the Elizabethan Era." *Early Modern Literary Studies*, vol. 4, no. 1, 1998, pp. 8.1-4. http://purl.oclc.org/emls/04-1/rev_fitt.html.

Accessed 3 April 2016.

Forbes, David. "How capitalism captured the mindfulness industry." *The Guardian*. 16 April 2009. https://www.theguardian.com/lifeandstyle /2019/apr/16/how-capitalism-captured-the-mindfulness-industry?CMP =share_btn_link. Accessed 26 April 2019.

Foshay, Raphael. *The Digital Nexus: Identity, Agency, and Political Engagement*. Athabasca University Press, 2016.

Friedman, Ted. "Apple's 1984: The Introduction of the Macintosh in the Cultural History of Personal Computers." Society for the History of Technology Convention, October 1997, Pasadena, California, Paper.

Fryer, Peter. *Staying Power: The History of Black People in Britain*. Pluto Press, 1984.

Fukuyama, Francis. "The End of History?" *The National Interest*, no. 16, 1989, pp. 3–18. *JSTOR*, www.jstor.org/stable/24027184. Accessed 7 June 2017.

Fuller, R. Buckminster. *Operating Manual for Spaceship Earth*. Estate of R. Buckminster Fuller, 2008.

Galloway Alexander. *"The Digital." Laruelle: Against the Digital*. University of Minnesota Press, 2014, pp. 51-72.

Galloway, Alexander R. *The Interface Effect*. Polity Press, 2012.

Gilbert, Marc Jason. "Next Stop: Silicon Valley: The Cold War, Vietnam and the Making of the California Economy." *What's Going On?: California and the Vietnam Era*. Edited by Marcia A. Eymann, Charles Wollenberg,

Oakland Museum of California, University of California Press, 2004, pp. 23-41.

Gaines, Steven. *Heroes and Villains: The True Story of The Beach Boys*. Hachette Books, 1995.

Gilbert, Pamela K. *Victorian Skin: Surface, Self, History*. Cornell University Press, 2019.

Gilmore, Paul. "The Telegraph in Black and White." *ELH*, vol. 69 no. 3, 2002, pp. 805-833. *Project MUSE*, doi:10.1353/elh.2002.0025.

Gilroy, Paul. *Between Camps: Nations, Cultures and the Allure of Race London*. Routledge, 2013.

---. *"Race and Racism 'In the Age of Obama.'" The Tenth Annual Eccles Centre for American Studies Plenary Lecture given at the British Association for American Studies Annual Conference. The British Library*, 2013, pp. 1-22, https://www.bl.uk/britishlibrary/~/media/bl/global/eccles%20cen tre/ec%20plenaries/baas-2013-gilroy.pdf. Accessed 26 June 2017.

---. "Race ends here." *Ethnic and Racial Studies*, vol. 21, no. 5, 1998, pp. 838-847, https://doi.org/10.1080/014198798329676. Accessed 12 April 2016.

Golub, Philip S. "A Taste of Blood in the Jungle Revisiting American Empire." *Soundings*, vol. 37, no. 1, 2007, pp. 66-79, doi:10.3898/136266207820465615. Accessed 15 May 2015.

Graeber, David. *The Utopia of Rules*. Melville House, 2015.

Graham, Stephen. "Digital Medieval." *Surveillance & Society*, vol. 9.no. 3, 201, pp. 321-327. https://doi.org/10.24908/ss.v9i3.4289. Accessed 28 March 2016.

Grow, Kory. "Charles Manson: How Cult Leader's Twisted Beatles Obsession Inspired Family Murders." *Rolling Stone*, 9 Aug 2017, https://www.rollingstone.com/culture/culture-features/charles-manson-how - cult - leaders-twisted-beatles-obsession-inspired-family-murders-107176/. Accessed 14 Jun 2019.

Hage, Ghassan. *Against Paranoid Nationalism: Searching for Hope in a Shrinking Society*. Pluto Press, 2003.

Haiven, Max. "Finance as Capital's Imagination? Reimagining Value and Culture in an Age of Fictitious Capital and Crisis." *Social Text 108*, vol. 29, no. 3, 2011, pp. 93-124, doi:10.1215/01642472-1299983.

Håkansson, Håkan. *Seeing the Word: John Dee and Renaissance Occultism*. Lund University, 2001.

Hall, Mirko M. "Dialectic Sonority: Walter Benjamin's Acoustics of Profane Illumination." *Telos*, vol. 152, 2010, 83-102, doi:10.3817/0910152083.

Halpern, Orit. *Beautiful Data: A History of Vision and Reason since 1945*. Duke University Press, 2015.

Haraway, Donna. "A Cyborg Manifesto." *Manifestly Haraway.* Minneapolis: University, 2016, pp. 3-90.

Harvey, David. *The New Imperialism.* Oxford University Press, 2003.

Harwood, John. *The Interface: IBM and the Transformation of Corporate Design, 1945–1976.* University of Minnesota Press, 2011.

Hawkesworth, Mary. *Embodied Power: Demystifying Disembodied Politics.* Routledge, 2016.

Hayles, N. Katherine. *How We Became Posthuman: Virtual Bodies in Cybernetics, Literature, and Informatics.* University of Chicago Press, 2008.

Hepler, Lauren. "Menial Tasks, Slurs and Swastikas: Many Black Workers at Tesla Say They Faced Racism." *The New York Times.* 30 Nov 2018, https://www.nytimes.com/2018/11/30/business/tesla-factory-racism.html. Accessed 26 Feb 2018.

Hern, Alex. "Roger McNamee: 'It's bigger than Facebook. This is a problem with the entire industry'". *The Observer.* 16 Feb 2019, https://www. theguardian.com/books/2019/feb/16/roger-mcnamee-zucked-waking-up -to-the- facebook - catastrophe - interview?CMP=Share_iOSApp_Other. Accessed 24 Feb 2019.

Hight, Christopher. "Stereo Types: The Operation of Sound in the Production of Racial Identity." *Leonardo*, vol. 36, no. 1, 2003, pp. 13–17. *JSTOR*, www.jstor.org/stable/1577272. Accessed 14 Oct 2017.

Hill, Zahara. "Charles Manson's Failed Attempts to Start a Race War." *Ebony.* 20 Nov 2017. https://www.ebony.com/news/helter-skelter-charles-manson-race-war-failed/. Accessed 13 Jun 2019.

Holston, James and Arjun Appadurai. "Introduction." *Cities and Citizenship.* Duke University Press, 1999.

"Is Asperger's syndrome the next stage of human evolution?: Tony Attwood | Australian Story." ABC News Australia. *YouTube.* https://www.you tube.com/watch?v=vdQDvLXLqiM. Accessed 2 May 2019.

Isaacson, Walter. "The Real Leadership Lessons of Steve Jobs." *Harvard Business Review.* April 2012. https://hbr.org/2012/04/the-real-leadership-lessons-of-steve-jobs. Accessed 20 April 2019.

Iton, Richard. *In Search of the Black Fantastic: Politics and Popular Culture in the Post-Civil Rights Era: Politics and Popular Culture in the Post-Civil Rights Era.* Oxford University Press, 2008.

"It's Time to Divest from Fossil Fuels." *350.org.* 27 Oct 2016, https://www.ecowatch.com/nobel-prize-climate-change-2066167963.html. Accessed 12 May 2019.

Jahme, Carole. "The fantastic Dr Dee: angels, magic and the birth of modern science." The Guardian. 25 June 2012. https://www.theguardian. Com

/science / 2012 / jun / 25 / fantastic – dr – dee – birth – modern - science. Accessed 26 July 2017.

Jandri´c, Petar. "From the Electronic Frontier to the Anthropocene." *Learning in the Age of Digital Reason*. Sense, 2017. https://doi.org/10.1007/9789 463510776_005. Accessed 2 March 2019.

Jandrić, Petar, and Ana Kuzmanić. "Digital Postcolonialism." *IADIS International Journal on WWW/Internet*, vol. 13. no. 2, 2016, pp. 34-51. http://www.iadisportal.org/ijwi/papers/2015131203.pdf. Accessed 10 Jan 2017.

Jean Baudrillard, Jean. "Simulacra and Simulations." *Jean Baudrillard, Selected Writings*. Edited by Mark Poster, Stanford University Press, 1988, pp. 166-184.

"John Lennon Interview: Playboy 1980." *The Beatles Ultimate Experience*, n.d. http://www.beatlesinterviews.org/db1980.jlpb.beatles.html. Accessed 15 Jun 2019.

Johnson, Poe. "Racial Technologies in the Time of Black Cyborgnetic Consciousness." *The Routledge Companion to Biology in Art and Architecture*. Edited by Charissa N. Terranova & Meredith Tromble, Routledge, 2017, pp. 368-384.

Kater, Michael H. "Forbidden Fruit? Jazz in the Third Reich." *The American Historical Review*, vol. 94, no. 1, 1989, pp. 11–43. JSTOR. www.jstor.org/stable/1862076. Accessed 2 Feb 2018.

Kaufman, Amy S. "Our Future Is Our Past: Corporate Medievalism in Dystopian Fiction." *Studies in Medievalism XXII: Corporate Medievalism II*, edited by Karl Fugelso, Boydell & Brewer, 2013, pp. 11–20. *JSTOR*, www.jstor.org/stable/10.7722/j.ctt2tt1q7. Accessed 6 May 2017.

Keller, Bill. "Henry Luce, the Editor in Chief." *New York Times Sunday Book Review*. 22 Apr 2010. https://www.nytimes.com/2010/04/25/books/ review/Keller-t.html. Accessed 11 Feb 2018.

Khalil, Yousef. "Neoliberalism and the Failure of the Arab Spring." *New Politics*. Vol. 15. No.3, 2015, pp 77-82, https://newpol.org/issue_post/neolibe ralis m-and-failure-arab-spring/. Accessed 28 Dec 2015.

Kirk, Andrew. "Appropriating Technology: The Whole Earth Catalog and Counterculture Environmental Politics." *Environmental History*, vol. 6, no. 3, 2001, pp. 374–394. JSTOR, www.jstor.org/stable/3985660. Accessed 4 Apr 2019.

Kodat, Catherine Gunther. "Conversing with Ourselves: Canon, Freedom, Jazz." *American Quarterly*, vol. 55, no. 1, 2003, pp. 1–28. *JSTOR*, www.jstor.org/stable/30041955. Accessed 12 Nov 2018.

Kow, Simon. *China in Early Enlightenment Political Thought*. London: Taylor & Francis, 2016.

LaGrandeur, Kevin. *Androids and Intelligent Networks in Early Modern Literature and Culture Artificial Slaves.* Routledge 2013.

Kreiss, Daniel. "Appropriating the Master's Tools: Sun Ra, the Black Panthers, and Black Consciousness, 1952-1973." *Black Music Research Journal*, vol. 28, no. 1, 2008, pp. 57–81. JSTOR, www.jstor.org/stable/25433794. Accessed 17 Jun 2019.

Kruse, Michael. "How Vietnam Became Donald Trump's Forever War." *Politico*, 26 Feb 2019, https://www.politico.com/magazine/story/2019/02/26/vietnam -donald-trump-forever-war-225210. Accessed 13 June 2019.

Leslie, Esther. "Those in Glass Houses Laugh." 27 September 2016. *Verso Blog.* https://www.versobooks.com/blogs/2855-those-in-glass-houses-laugh. Accessed 14 Nov 2018.

Leslie, Thomas W. "Energetic Geometries: the Dymaxion Map and the Skin/Structure Fusion of Buckminster Fuller's Geodesics." *Architectural Research Quarterly*, vol. 5, no. 2, 2001, pp. 161–170., doi:10.1017/S135913550100118X.

"Letters to Walter Benjamin." *Aesthetics and Politics*, edited by Fredric Jameson. London: New Left Books, 1977, pp.110-133.

Levy, Steven. "One More Thing Inside Apple's Insanely Great (or Just Insane) New Mothership." WIRED. 5 May 2017, https://www.wired.com/2017/05/apple-park-new-silicon-valley-campus/. Accessed 18 July 2017.

Lin, Tan. "Disco as Operating System, Part One." *Criticism*, Winter 2008, Vol. 50, No. 1, pp. 83–100. https://digitalcommons.wayne.edu/criticism/vol50/iss1/5. Accessed 15 April 2019.

Lindroos, Kia. "Aesthetic Political Thought: Benjamin and Marker Revisited." *Alternatives: Global, Local, Political*, vol. 28, no. 2, 2003, pp. 233–252. *JSTOR*, www.jstor.org/stable/40645077. Accessed 11 Dec 2015.

Linebaugh, Peter and Marcus Rediker. *The Many-Headed Hydra: Sailors, Slaves, Commoners, and the Hidden History of the Revolutionary Atlantic.* Beacon Press, 2013.

Lock, Graham. *Blutopia: Visions of the Future and Revisions of the Past in the Work of Sun Ra, Duke Ellington, and Anthony Braxton.* Duke University Press, 1999.

Louv, Jason. "John Dee was the 16th century's real-life Gandalf." *Boing Boing.* 19 Feb 2015. https://boingboing.net/2015/02/19/john-dee-was-the-real-life-mer.html. Accessed 7 Sept 2017.

---. *The Angelic Reformation: John Dee, Enochian Magick & the Occult Roots of Empire.* Ultraculture Incorporated, 2015.

Louw, Eric P. The Rise, Fall, and Legacy of Apartheid. Greenwood Publishing Group, 2004.

Lowe, Lisa. *The Intimacies of the Four Continents.* Duke University Press, 2015.

Luce, Henry R. "The American Century." *Diplomatic History*, vol. 23, no. 2, April 1999, Pages 159–171, https://doi.org/10.1111/1467-7709.00161.

Lütticken, Sven. "Hito Steyerl: Postcinematic Essays After the Future." *Too Much World: The Films of Hito Steyerl.* Edited by Nick Aikens, Van Abbemuseum/Sternberg Press, 2014, pp. 45-62.

Magubane, Bernard. *The Making of a Racist State: British Imperialism and the Union of South Africa, 1875-1910.* Africa World Press, 1996. Marks, Laura U."*Monad, Database, Remix*: Manners of Unfolding in *The Last Angel of History*." *Black Camera*, vol. 6 no. 2, 2015, pp. 112-134. *Project MUSE*, muse.jhu.edu/article/583175. Accessed 6 Sept 2017.

Annie McClanahan. "INVESTING IN THE FUTURE: Late capitalism's end of history" *Journal of Cultural Economy*, Vol 6, No. 1, pp. 78-93, https://doi.org/10.1080/17530350.2012.745442. Accessed 18 March 2016.

McClintock, Anne. "Soft Soaping empire: commodity racism and imperial advertising." *Travellers' Tales: Narratives of Home and Displacement.* Edited by Jon Bird, Barry Curtis, Melinda Mash, Tim Putnam, George Robertson, Lisa Tickner, London: Routledge, 2005, pp. 131-154.

McElroy, Erin. "Postsocialism and the Tech Boom 2.0: techno-utopics of racial/spatial dispossession." *Social Identities.* 2017, pp. 1-16. https://doi.org/10.1080/13504630.2017.1321718.

McFarland Philip James. *Constellation: Friedrich Nietzsche and Walter Benjamin in the Now-Time of History.* Fordham University Press, 2012.

McGuire, Anne. "De-regulating Disorder: On the Rise of the Spectrum as a Neoliberal Metric of Human Value." *Journal of Literary & Cultural Disability Studies*, vol. 11 no. 4, 2017, pp. 403-421. *Project MUSE*, muse.jhu.edu/article/677601. Accessed 1 May 2019.

Michaels, Walter Benn. "Let Them Eat Diversity." *Jacobin.* https://jacobinmag.com/2011/01/let-them-eat-diversity. Accessed 25 May 2016.

Mirzoeff, Nicholas. "It's Not the Anthropocene, It's the White Supremacy Scene; or, the Geological Color Line," *After Extinction*, edited by Grusin, R. Minnesota: University of Minnesota Press, 2018, pp. 123-150.

Morningstar, Corey. "The Manufacturing of Greta Thunberg – For Consent: The Political Economy of The Non-Profit Industrial Complex [Act I]." *Wrong Kind of Green.* 17 Jan 2019. http://www.wrongkindofgreen. Org / 2019 / 01 / 17 / the - manufacturing-of-greta-thunberg-for-consent-the-political-economy-of-the-non-profit-industrial-complex/. Accessed 10 May 2019.

Mörtenböck, Peter. "Global Informality: Bottom-Up Trade and Transnational Realignments." *Informal Market Worlds Reader: The Architecture of*

Economic Pressure, edited by Peter Mörtenböeck, Helge Mooshammer, Teddy Cruz, Fonna Forman, nai010 Publishers, 2015, pp. 105-118.

Mosco, Vincent. *Becoming Digital: Toward a Post-Internet Society.* Emerald Group Publishing, 2017.

Nakamura, Lisa. *Digitizing Race: Visual Cultures of the Internet.* University of Minnesota Press, 2008.

Neuding, Paulina. "Self-Harm Versus the Greater Good: Greta Thunberg and Child Activism." / *Quillete.* 23 Apr 2019, https://quillette.com/2019/ 04 / 23 / self-harm – versus – the – greater – good-greta-thunberg-and-child-activism. Accessed 2 May 2019.

"Never Learn Not To Love The Beach Boys." *Genius,* n.d. https://genius.com/ The-beach-boys-never-learn-not-to-love-lyrics. Accessed 15 June 2019.

Newman, Bruce. "Steve Jobs, visionary leader of Apple, dies at 56." *The Seattle Times.* Oct 5 2011, https://www.seattletimes.com/nation-world/steve-jobs-visionary-leader-of-apple-dies-at-56/. Accessed 2 Apr 2019.

Ngai, Mae M. "Trouble on the Rand: The Chinese Question in South Africa and the Apogee of White Settlerism." *International Labor and Working-Class History,* vol. 91, 2017, pp. 59–78., doi:10.1017/S014754791600 0326.

"Night Fever Bee Gees." *Genius,* n.d. https://genius.com/Bee-gees-night-fever-lyrics. Accessed 2 May 2019.

Noys, Benjamin. *Malign Velocities: Accelerationism and Capitalism.* Zero Books, 2014.

Oppenheimer, Mark. Knocking on Heaven's Door: American Religion in the Age of Counterculture. Yale University Press, 2003.

Orlando, Jordan. "The Accidental Perfection of the Beatles' White Album." *The New Yorker,*

10 Nov 2018, https://www.newyorker.com/culture/cultural-comment/the-accidental-perfection-of-the-beatles-white-album. Accessed 14 Jun 2019.

Packer, George. "The Other France." *The New Yorker.* 31 Aug 2015, https:// www.newyorker.com/magazine/2015/08/31/the-other-france. Accessed 26 Dec 2015.

Parikka, Jussi. "On Media Meteorology." *Machinology.* 31 May 2017, https://jussiparikka.net/2017/05/31/on-media-meteorology/. Accessed 6 Sept 2017.

---. "Cultural techniques of cognitive capitalism: Metaprogramming and the labour of code." *Cultural Studies Review.* vol.20, no, 1, March 2014, pp. 30-52. https://doi.org/10.5130/csr.v20i1.3831.

Park, Lisa Sun-Hee, and David N. Pellow. "Racial Formation, Environmental Racism, and the Emergence of Silicon Valley." *Ethnicities,* vol. 4, no. 3, Sept. 2004, pp. 403–424, doi:10.1177/1468796804045241.

Pasquinelli, Matteo. "The Automaton of the Anthropocene: On Carbosilicon Machines and Cyberfossil Capital." *South Atlantic Quarterly*. Vol. 116, no. 2, April 2004, pp. 311–326, doi: https://doi.org/10.1215/00382876-3829423

---. "Introduction" Alleys of Your Mind: Augmented Intelligence and Its Traumas." Edited by Matteo Pasquinelli, Meson Press, 2015, pp. 7-22.

Peck, Janice. "Oprah Winfrey Cultural Icon of Mainstream (White) America." *The Colorblind Screen: Television in Post-Racial America.* Edited by Sarah Nielsen and Sarah E. Turner, NYU Press, 2014, pp. 83-107.

Pinto, Ana Teixeira and Anselm Franke. "The Post-Internet Condition." *ARTis On*, No. 4, 2016, pp 26-31, http://artison.letras.ulisboa.pt/index.php/ao/article/view/100. Accessed 10 Jan 2016.

Pinto, Ana. Teixeira "The Pigeon in the Machine: The Concept of Control in Behaviorism and Cybernetics." *Alleys of Your Mind: Augmented Intelligence and Its Trauma*, edited by Matteo Pasquinelli, Meson Press, 2015 pp. 23-36.

---. "I have no way of knowing if you can hear me." *Amalia Pica*. Edited by Ribas, João, Julie Rodrigues Widholm, and Amalia Pica, Museum of Contemporary Art Chicago, 2013, pp, 32-37.

Pohl, Stephen. "The Cybernetic Delirium of Norbert Wiener." *CTheory*, 30 Jan 1997, http://ctheory.net/ctheory_wp/the-cybernetic-delirium-of-norbert-wiener/. Accessed 25 Sep 2017.

Purser, Ron and David Loy. "Beyond McMindfulness." *The Huffington Post.* 31 Aug 2013. https://www.huffpost.com/entry/beyond-mcmindfulness_b_3519289. Accessed 17 Apr 2018.

Rabinow, Paul and Nikolas Rose. "Biopower Today". *BioSocietise*, vol. 1, issue 2, June 2006, pp. 195–217, https://doi.org/10.1017/S17458552060400 14. Accessed 19 May 2016.

Redclift, Victoria. "New racisms, new racial subjects? The neo-liberal moment and the racial landscape of contemporary Britain." *Ethnic and Racial Studies*, vol. 37, no. 4 2014, pp. 577-588. 03 Jan 2014, https://doi.org/10.1080/01419870.2014.857032. Accessed 27 Sep 2017.

Reitman, Meredith. "Uncovering the White Place: Whitewashing at Work." *Social & Cultural Geography*, vol. 7, no. 2, 2006, pp. 267-282. https://doi.org/10.1080/14649360600600692. Accessed 15 May 2015.

"Rename Milner Hall." Cross Rhodes. http://crossrhodestt.com/campaigns/milner-hall/.Accessed 11 Feb 2019.

"Revolution 9 by The Beatles." *Songfacts*, n.d. https://www.songfacts.com/facts/the-beatles/revolution-9. Accessed 3 Jun 2019.

Richter, Gerhard. *Walter Benjamin and the Corpus of Autobiography*. Wayne State University Press, 2002.

Ronell, Avital. *Finitude's Score: Essays for the End of the Millennium*. University of Nebraska Press, 1994.

Rosenfeld, Megan. "Encountering Werner Erhard". *The Washington Post.* 14 Apr 1979. https://www.washingtonpost.com/archive/lifestyle/1979/04/14/encountering-werner-erhard/00d942f2-1f67-4dc8-b385-cfadae 1437b5/?utm_term=.642c065329e7#comments. Accessed 8 May 2018.

Rosenwald, Michael, S. "Charles Manson's surreal summer with the Beach Boys: Group sex, dumpster diving and rock 'n' roll." *The Washington Post*, 20 Nov 2017, https://www.washingtonpost.com/news/retropolis/wp/2017/11/20/charles-mansons-surreal-summer-with-the-beach-boys-group-sex-dumpster-diving-and-rock-n-roll/?utm_term = .4100e295c5be. Accessed 15 Jun 2019.

Rouvroy, Antoinette. "Revitalizing Critique Against the Critical Sirens of Algorithmic Governmentality." *YouTube*, uploaded by Teo Cruz 5 Jun 2017, https://www.youtube.com/watch?v=ezd24NQoqCw&t=1540s.

Roy, Ananya, Stuart Schrader, and Emma Shaw Crane. "'The Anti-Poverty Hoax': Development, pacification, and the making of community in the global 1960s." *Cities: The International Journal of Urban Policy and Planning*, vol.44, Apr 2015, pp. 139-145, https://doi.org/10.1016/j.cities. 2014.07.005. Accessed 26 Sep 2017.

Rutsky, R. L."Walter Benjamin and the Dispersion of Cinema." *symploke*, vol. 15 no. 1, 2007, pp. 8-23. *Project MUSE*, doi:10.1353/sym.0.0017. Accessed 28 Dec. 2015.

Ryder, Robert G. "WALTER BENJAMIN'S SHELL-SHOCK: Sounding the acoustical unconscious." *New Review of Film and Television Studies*, vol. 5, no. 2, 2007, pp.135-155. https://doi.org/10.1080/174003007014 32811 Accessed 30 Dec 2015.

Salamon, Margaret Klein. "Leading the Public into Emergency Mode." *The Climate Mobilization*. https://www.theclimatemobilization.org/emergency-mode. Accessed 10 May 2019.

Sanson, Kevin. "Corresponding geographies: Remapping work and workplace in the age of digital media." *Television & New Media*, vol. 16, no. 8, 2004, pp. 1-18. https://doi.org/10.1177/1527476414559289. Accessed 10 May 2015.

Sanyal, Debarati. "Auschwitz as Allegory in *Night and Fog.*" Concentrationary Cinema: Aesthetics as Political Resistance in Alain Resnais's Night and Fog. Edited by Griselda Pollock and Max Silverman, Berghahn Books, 2012, pp. 152-182.

Shah, Nishat. "Identity and Identification: The Individual in the Time of Networked Governance.*" The Socio-legal Review*, vol. 11, no. 2, 2015, pp. 22-40. http://www.sociolegalreview.com/wp-content/uploads/2015/12/Identity-and-Identification-the-Individual-in-the-Time-of-Networked-Governance.pdf. Accessed 6 Sept 2017.

Shapiro, Michael J. "Every Move You Make: Bodies, Surveillance, and Media." *Social Text. 23*, vol.2, no. 83 ,2005, pp. 21-34. https://doi.org/10.1215/01642472-23-2_83-21. Accessed 28 March 2016.

Sharma, Sanjay. "Black Twitter?: Racial Hashtags, Networks and Contagion." *new formations: a journal of culture/theory/politics*, vol. 78, 2013, pp. 46-64. *Project MUSE*, muse.jhu.edu/article/522093. Accessed 28 May 2016.

Shaviro, Steven. "Thinking Blind." Journal of the Fantastic in the Arts, vol. 25, no. 2/3 (91), 2014, pp. 314-331, https://www.jstor.org/stable/i24353022. Accessed 4 May 2019.

---. "Panpsychism And/Or Eliminativism." 4 Oct 2011. http://www.shaviro.com/Blog/?p=1012. Accessed 4 May 2019.

---. *Post Cinematic Affect*. John Hunt Publishing, 2010.

Sherman, William Howard. *John Dee: The politics of reading and writing in the English Renaissance.* University of Massachusetts Press, 1997.

Silberman, Steve. *Neurotribes: The Legacy of Autism and the Future of Neurodiversity.* Penguin, 2015.

Silbey, David J. *The Boxer Rebellion and the Great Game in China: A History.* Farrar, Straus and Giroux, 2012.

Sims, Alexandra. "Sweden on target to run entirely on renewable energy by 2040." *The Independent.* 26 Oct 2016. https://www.independent.co.uk/news/world/europe/sweden-renewable-energy-target-2040-country-on-track-a7381686.html. Accessed 10 May 2019.

Singh, Nikhil Pal. *Race and America's Long War.* University of California Press, 2017.

Spikins, Penny. "What Role Did Autism Play in Human Evolution?". Sapiens. 9 May 2017. https://www.sapiens.org/evolution/autism-human-evolution/. Accessed 9 May 2019

Staples, Brent. "Music of the Spheres." *The New York Times*, 17 Aug 1997, https://archive.nytimes.com/www.nytimes.com/books/97/08/17/reviews/970817.17staplet.html?module=inline. Accessed 18 Jun 2019.

Steinhorn, Leonard and Barbara Diggs-Brown. *By the Color of Our Skin.* Dutton, 1999.

Steyerl, Hito. "Digital Debris: Spam and Scam." *October,* vol.138, 2011, pp. 70-80. https://doi.org/10.1162/OCTO_a_00067. Accessed 11 Dec 2015.

---. "In Free Fall: A Thought Experiment on Vertical Perspective." *The Wretched of the Screen.* E-flux and Sternberg Press, 2012.

---. "Too Much World: Is the Internet Dead?" *e-flux journal*, vol. 49, Nov 2013, https://www.e-flux.com/journal/49/60004/too-much-world-is-the-internet-dead/. Accessed 19 March 2016.

Street, Paul. *Barack Obama and the Future of American Politics.* Paradigm, 2009.

Sutherland, Tonia. "Making a Killing: On Race, Ritual, and (Re)Membering in Digital Culture." Preservation, Digital Technology & Culture, vol. 46, no. 1, 2017, pp. 32-40, https://doi.org/10.1515/pdtc-2017-0025. Accessed 29 Sep 2017.

Tao, Zhijian. *Drawing the Dragon: Western European Reinvention of China.* Peter Lang, 2009.

Terranova, Tiziana. *Network Culture: Politics for the Information Age.* Pluto, 2004.

Teskey, Gordon. *Allegory and Violence.* Ithaca: Cornell University Press, 1996.

"The Beatles - Revolution Lyrics." *MetroLyrics*, n.d. http://www.metrolyrics.com/revolution-lyrics-beatles.html. Accessed 15 Jun 2019.

"The U.S. Military and Oil." *Union of Concerned Scientists.* n.d. https://www.ucsusa.org/clean_vehicles/smart-transportation-solutions/us-military-oil-use.html. Accessed 8 May 2019.

Thunberg, Greta. "'Our house is on fire': Greta Thunberg, 16, urges leaders to act on climate." *The Guardian.* 25 Jan 2019. https://www.theguardian.com/environment/2019/jan/25/our-house-is-on-fire-greta-thunberg16-urges-leaders-to-act-on-climate. Accessed 2 May 2019.

Thunberg, Greta. "You did not act in time': Greta Thunberg's full speech to MPs." *The Guardian.* 23 Apr 2019. https://www.theguardian.com/environment /2019/apr/23/greta-thunberg-full-speech-to-mps-you-did-not-act-in-time. Accessed 3 May 2019.

Tiqqun. *Introduction to Civil War.* Translated by Alexander R. Galloway and Jason E. Smith, MIT Press, 2010.

Titley, Gavan. "No apologies for cross-posting: European trans-media space and the digital circuitries of racism." *Crossings: Journal of Migration & Culture*, vol. 5, no 1, March 2014, pp. 41-55. https://doi.org/10.1386/cjmc.5.1.41_1. Accessed 27 Sept 2017.

Trotter, David. "Mobility, network, message: spy fiction and film in the long 1930s." *Source Information*, vol. 57, no. 3, October 2015, pp. 1-21. https://doi.org/10.1111/criq.12213. Accessed 12 Jan 2019.

Turner, Fred. "Where the counterculture met the new economy: The WELL and the origins of virtual community." *Technology and Culture*, vol. 46, no.3, 2005, pp. 485-512, 10.1353/tech.2005.0154. Accessed 15 Apr 2019.

Turner, Fred. *From Counterculture to Cyberculture: Stewart Brand, the Whole Earth Network, and the Rise of Digital Utopianism.* University of Chicago Press, 2010.

Turner, Henry S. "Corporations: Humanism and Elizabethan Political Economy." *Mercantilism Reimagined: Political Economy in Early Modern Britain and Its Empire.* Oxford University Press, 2014, pp. 153-76.

---. *The Corporate Commonwealth: Pluralism and Political Fictions in England, 1516-1651.* University of Chicago Press, 2016.

---. *Early Modern Theatricality.* Oxford University Press, 2013.

Turing, Alan M. "Computing Machinery and Intelligence." *Mind 49*: 433-466. *Mind*, vol. LIX, no. 236, October 1950, pp. 433–460, https://doi.org/10.1093/mind/LIX.236.433. Accessed 14 Jan 2019.

Turpin, Etienne. "The Same River Twice: Torrential Formations of the Anthropocene." *YouTube*, uploaded by Teo Cruz, 5 Jun 2017, https://www.youtube.com/watch?v=2ZLlSLEDIoU.

Verrall, Krys. "Art and Urban Renewal: MoMa's New City Exhibition and Halifax's Uniacke Square." *The Sixties: Passion, Politics, and Style.* Edited by Dimitry Anastakis. McGill-Queen's Press, 2014, pp. 145-166.

Vishmidt, Marina. "The cultural logic of criticality." *Journal of Visual Art Practice*, vol. 7, no. 3, 2008, pp. 253-269, doi: 10.1386/jvap.7.3.253_1. Accessed 28 Sep 2017.

Wald, Gayle. "Soul Vibrations: Black Music and Black Freedom in Sound and Space." *American Quarterly*, vol. 63 no. 3, 2011, pp. 673-696. *Project MUSE*, doi:10.1353/aq.2011.0048. Accessed 10 Mar 2019.

Walsh, Mike. "Sun Ra: Stranger from Outer Space." Mission Creep, http://www.missioncreep.com/mw/sunra.html. Accessed 17 Jun 2019.

Walter, Damien. "The ancient book of wisdom at the heart of every computer." *The Guardian.* 21 March 2014. https://www.theguardian.com/books/2014/mar/21/ancient-book-wisdom-i-ching-computer-binary-code. Accessed 20 Mar 2019.

Wark, McKenzie. "Escape from the Dual Empire." *Rhizomes.* 6 Sept 2017. http://www.rhizomes.net/issue6/wark.htm. Accessed 15 Sept 2017.

---. "On Hito Steyerl" *Public Seminar*, 14 Aug 2015, http://www.public seminar.org/2015/04/on-hito-steyerl/. Accessed 5 Dec. 2015.

---. "The Vectoralist Class*." e-flux journal Supercommunity*, 29 Aug 2015, http://supercommunity.e-flux.com/texts/the-vectoralist-class/. Accessed 28 Dec 2015.

--. *A Hacker Manifesto.* Harvard University Press, 2004.

Web 28 Dec. 2015.

"We Don't Have Time". YouTube, uploaded by *We Don't Have Time*, 6 Apr 2018, https://www.youtube.com/watch?time_continue=1&v=pEhQ8gqRxbc.

Weiner, Norbert. "Father of Cybernetics Norbert Wiener's Letter to UAW President Walter Reuther." *libcom.org*, https://libcom.org/history/father-cybernetics-norbert-wieners-letter-uaw-president-walter-reuther. Accessed 25 Sep 2017.

Williamson, Ben. "Educating Silicon Valley: Corporate Education Reform and the Reproduction of the Techno-Economic Revolution." *Review of*

Education, Pedagogy, and Cultural Studies vol. 39, no.3, 2017, pp. 265–288. https://doi.org/10.1080/10714413.2017.1326274

Winant, Howard. "The Dark Matter: Race and Racism in the 21st Century." *Critical Sociology*, vol. 41, no. 2, Mar. 2015, pp. 313–324, doi:10.1177/0896920513501353. Accessed 27 March 2016.

Wipplinger, Jonathan. "The Aural Shock of Modernity: Weimar's Experience of Jazz." The Germanic Review: Literature, Culture, Theory, *Taylor and Francis Journals*, vol. 82, no. 4, 2007: pp. 299–320, doi 10.3200 /GERR.82.4.299-320. Accessed 24 Nov 2015

Wolfe, Patrick. "Settler colonialism and the elimination of the native." *Journal of Genocide Research*, vol.8, no. 4, 2006, pp.387–409, https://doi.org/10.1080/14623520601056240. Accessed 3 June 2015.

Woolley, Benjamin. *The Queen's Conjurer: The Science and Magic of Dr. John Dee, Advisor to Queen Elizabeth I*. Macmillan, 2002.

Wyly Elvin K., and Jatinder K. Dhillon. "Planetary Kantsaywhere Cognitive capitalist universities and accumulation by cognitive dispossession" *City*, vol. 22, no. 1, 2008, pp. 130-151. https://doi.org/10.1080/136048 13.2018.1434307. Accessed 21 Nov 2018.

Yancy George and Judith Butler. "What's Wrong With 'All Lives Matter'?" *The New York Times*. 12 Jan 2015, www.opinator.blog.nytimes.com. Accessed 28 Sep 2017.

Yancy, George and Paul Gilroy. "What 'Black Lives' Means in Britain." *The New York Times*. 1 Oct 2015, https://opinionator.blogs.nytimes.com /2015/10/01/paul-gilroy-what-black-means-in-britain/ Accessed 3 May 2016.

Young, Stephen, Alasdair Pinkerton & Klaus Dodds. "The word on the street: Rumor, "race" and the anticipation of urban unrest." *Political Geography*, vol. 38, 2014, pp. 57-67, https://doi.org/10.1016/j.polgeo. 2013.11.001. Accessed 23 Oct 2017.

Zhou, Yongming. *Historicizing online politics: Telegraphy, the Internet, and political participation in China*. Stanford University Press, 2006.

Zuberi, Nabeel. "The transmolecularization of [black] folk: Space is the Place, Sun Ra and Afrofuturism." *Off the planet: Music, sound and science fiction cinema*, vol. 7, 2004, pp. 77-95.

Zylinska, Joanna. *The End of Man: A Feminist Counterapocalypse*. University of Minnesota Press, 2018.